AI HACKED 2

To my friend Jackie Morey,
Merry Christmas and
Happy New Year!
God bless you all!

Elvar Simon
December 2020

AI HACKED 2

REIMAGINE THE FUTURE

ELZAR SIMON

E. SIMON PUBLISHING

Praises for AI Hacked 2

"Elzar Simon has adeptly demonstrated us that the future is now and we may not be ready for it. AI is not solely futuristic. AI is already embedded in nearly everything and the growth of AI is inevitable. He expertly guides us in imagining this future and our role, as humans, in it. Rather than 'us versus machine,' AI Hacked 2 outlines our unique human qualities that cannot be replicated. As a society, cultivating these unique human qualities and reimagining our relationship with work and purpose are more important than ever. Elzar Simon provides an approachable and defensible catalyst for this essential conversation."

--**Nancy E. Hill**, Ph.D., Developmental Psychologist and the Charles Bigelow Professor of Education, Harvard University

"Comprehensive review of current AI capabilities and very thoughtful perspective on most pressing issues around AI - opportunities, threats, applications, and policies. Highly useful advice and blueprints on how to maintain skills, persistence, and relevance as AI transforms every part of our lives. All in an accessible narration that carries the reader across topics."

--**Natalija Jovanovic**, MIT PhD, Chief Digital Officer, Sanofi Pasteur, Boston, MA United States

"Just finished reading Elzar Simon's follow up to AI Hacked, AI Hacked 2. It was worth the wait. It takes us further and deeper into the world occupied by AI and details the benefits and risks that these tools provide. Undoubtedly there will be challenges to damage limitation where the unscrupulous might tread, but overall the benefits are potentially enormous. They are some useful warnings on the socio-economic effects that might yet materialise, maybe needing legislation to contain. I recommend the book for its coverage and clarity of explanation."

--**Maurice Winn**, Chairman of Info Layer Ltd. A UK based Business Intelligence Company

"When I was a younger man, artificial intelligence existed only in the realm of science fiction. Today AI promises (or threatens) drastic change on a scale that even the best sci-fi writers couldn't imagine. In this insightful book, futurist Elzar Simon paints a balanced picture of how AI is emerging that will help you understand and prepare for the transformation we can expect within our lifetimes."

--**Leigh Berrell**, Chief Information Officer (and 40-year technology veteran), Australia

"AI is about to engulf us with a ferocity akin to a tsunami. Read this book and prepare yourself for the onslaught!"

--**Jan Slawinski**, Terminal Operating Systems Consultant, Holbeach, U.K.

First paperback edition, June 2020

ISBN-13: 978-1-7351037-1-6 (Paperback)
ISBN-13: 978-1-7351037-0-9 (eBook)
ISBN-13: 978-1-7351037-2-3 (Audiobook)

Cover design by Stephen Simon
Printed in the United States of America

Published by E. Simon Publishing
https://www.esimonpublishing.com/

This book is lovingly dedicated to my wife, Dr. Elisa Simon, and to my children: Joshua John Simon, Elisha Paul Simon, and Ysabel Grace Simon.

Special thanks to the following people who have been a part of this journey:

My wife, Dr. Elisa Simon, for all the love and support
My daughter, Ysabel Simon, for managing this book project
My two sons, Joshua Simon and Elisha Simon,
for being a constant source of encouragement.

Melissa Moran, for her outstanding editing work,
Stephen Simon, for this compelling book cover.

For my book endorsers, Dr. Nancy Hill of Harvard University,
Dr. Natalija Jovanovic of MIT and Sanofi Pasteur,
Maurice Winn, Chairman of Info Layer Ltd of U.K.,
Janusz Slawinski, ports consultant based in U.K.,
and Leigh Berrell, CIO based in Australia.

Deo Gratias!

Table of Contents

Introduction

A fierce battalion has just crossed the Rubicon. The sound of a million ironclad footsteps in strict cadence reverberates across the land. The army has breached our frontiers with steely determination. Though this mighty infantry is our own creation, it is sweeping across our dominion with no reservation. Its tireless soldiers have no heart and no soul.

This restless force has a name: Artificial Intelligence, often referred to as AI. Some call its troops intelligent machines.

Is AI our enemy or is it mankind's much needed reinforcement?

As this AI army of machines makes its unrelenting advance, many are reasonably concerned while others welcome its arrival. This epochal meeting of man and machines is driving all of humanity into uncharted territory, a place where no human being has ever been before.

Throughout history, technology has been a tool of efficiency to mankind, not an autonomous entity that can think and act like humans. Until today.

With the advent of intelligent machines, the competition between AI and human laborers has not only become increasingly stiff, some believe that the winner is predictable. Machines do not get sick. They are emotionless, efficient, and consistent. They do not take breaks or vacations. They don't need medical benefits.

Just how good is this AI that stands as a competitor to Earth's most intelligent creatures?

An ancient Chinese game called Go, which dates back some 4,000 years and mastered by some of the most brilliant scholars, was recently subjugated by an AI machine: Google DeepMind's AlphaGo[1].

A masterful victory indeed, considering that mathematicians had estimated that the longest game length or possible board positions in Go is one (1) followed by 170 zeroes—more than the number of atoms in the universe.[2]

In the last quarter of 2019, the South Korean Go world champion, Lee Se-dol, announced his retirement because AI "cannot be defeated."[3]

This was not the first time that AI had baffled us at our own game. Eight years before Lee Se-dol's retirement, IBM Watson [4] humbled exceptional contestants in the celebrated TV game show *Jeopardy!*[5]

If those two events are a measure of AI's potential, then humanity is clearly up against some tough competition. As the rate of research and development in the field of AI is not slowing down, it is just a matter of time before we see the full impact of this irrepressible revolution.

In the next ten years, automation will take over all major industries. In transport, driverless cars, buses, trucks, and trains can spell the loss of millions of jobs.

In manufacturing, Oxford Economics has predicted that robots will potentially displace 20 million factory jobs by 2030.[6]

In offices as well, robotic process automation (RPA) solutions are bound to take over thousands, if not millions, of clerical jobs. Forrester, a prominent market research company, has predicted that cubicle-related jobs (clerical in nature) will be automated by 2030, eliminating about 20 million jobs. Another 29.9 million location-based workers, such as grocery store clerks, will be replaced by automation by 2030.[7]

Robot chefs, robot waiters, robot cashiers, robot concierges, robot bell captains, robot call center workers, robot attendants, robot hospital assistants, and robot patrol guards can potentially turn millions of people into unemployment statistics.

As labor becomes more efficient and less expensive, the economic scale will tip to the side of the wealthy. The rich will become richer, the poor will become poorer.

Furthermore, there is growing concern over the loss of privacy and liberty. It is no longer a secret that China has led the way in widespread use of facial recognition technology. France has announced that they will use facial recognition as their national ID system. Even Dubai is seriously considering microchip implants that can contain personal information like medical history, passwords, among others. That also means that every individual can be trackable.

More alarming is that authoritarian governments will be able to use AI to not only suppress the freedom of their own people, but also extend their surveillance and control beyond their borders. This may mean that those dictatorial regimes would have the capability to retaliate against other countries they deem as enemies by shutting down their key infrastructure.

Meanwhile, others—the visionaries or proponents of AI—have valid reasons to be convinced that this highly advanced technology will help us find solutions to the most challenging problems of mankind.

Artificial intelligence can accelerate drug discovery to cure some of the most devastating diseases known to man—like the coronavirus that caused the recent pandemic.

Robot-assisted surgery is now making it possible to perform minimally invasive and highly delicate operations that were once considered too risky due to surgeons' natural hand tremors.

Medical diagnostics powered by AI will soon make diagnosis much quicker and more accurate—crucial in medical cases where time is of the essence.

Mind-controlled prosthetics promises to deliver many disabled people from the bondage of immobility.

Personalized and interactive healthcare provided by internet-based AI platforms like Babylon Health[8] will make health service accessible and affordable to the public—around the globe.

The application of AI in banking and finance is already enhancing efficient conduct of business and personal transactions—loan processing, opening of bank accounts, among others. Even on-demand insurance made possible by AI will be beneficial to many.

Homes can be protected by AI-enabled security systems. Automated management of power and water consumption will result in savings and will contribute to the world's sustainability efforts.

The prospect of interconnecting the human brain to AI-enabled machines can deliver endless possibilities. Besides, don't we humans need robust augmentation to our brain power? Our memory is limited and at times faulty. The speed of our mental computational capabilities is no match to that of AI engines.

AI technology used for autonomous vehicles—despite the feared takeover of millions of jobs—can minimize road accidents, significantly reduce traffic congestion and carbon emissions, and provide better mobility for the elderly and disabled. Self-driving cars can even be used to deliver groceries and online orders during a period of mandatory lockdown and social distancing.

Furthermore, even robots that are deemed to have the potential to take over labor-intensive jobs can improve overall safety in the workplace as they can perform tasks that may be hazardous to humans.

As we speak, the early stages of AI metamorphosis are unfolding right before our very eyes. By 2040 or 2050, AI will have reached a level of maturity that will unleash its true potential.

Will AI be our doom or our deliverance?

While a lot of people are fearful of AI taking over jobs, some seers speculate that the elimination of human labor can be the nirvana that mankind is waiting for.

What about AI overtaking human intelligence, though?

A project called OpenCog[9] aims to build an open source network of AI algorithms that will enable AI to *improve* AI. In simple terms, an ecosystem of AI functions is foreseen that will make AI smarter and smarter to the point that it will achieve what is called *Artificial General Intelligence* (AGI), and subsequently attain *Artificial Super Intelligence* (ASI). AI will then be more intelligent than human beings.

A lot of people are very uncomfortable with the idea of machines becoming intellectually superior to all of mankind. But is that necessarily a cause for alarm—especially if we can all benefit from their super intelligence? On the other hand, can machines turn into ruthless masters of the human race?

While a tech-savvy few are gaining insights on AI's rapid development, many remain oblivious to its reality. Those who do stay updated on AI, however, insist that public debates have to happen now. Not five years from now. They should take place *now*—before it is too late.

Why?

For one, because there is a need for legislation and regulations around the use of robots and AI—which will take time.

Furthermore, retraining and upskilling of the workforce will take time as well.

Finally, the recent coronavirus pandemic will likely accelerate the development and use of AI and robots across many industries.

True, there is no assurance that arguments for and against AI can stop this global movement, but the general public has the right to know. Many might not like what they will hear, but they nonetheless have the right to such information. In the same way that, although a patient might not like the diagnosis, that person still has the right to know about his or her medical condition, including options that may be available.

In the case of AI, the opinions are polarized, the diagnoses are conflicting, and the options are binary—ignore it or do something about it.

Arresting its development is not a viable choice. Restrict AI research in one region and another country will gladly advance AI development. It is a global evolution that no one person or group can stop. The reality is that countries and companies are racing to get ahead of the AI pack.

The country that dominates AI and reinforces its military capability with robotics and AI technology will become the next superpower.

It is not a remote possibility that we may find ourselves in another Cold War era with a shade of ultramodern feudalism. AI can be an instrument of global division: polarization between two superpowers, and the rich lording over the poor.

We now have reached a tipping point—a point of no return.

There is no turning back, as AI's troops are forging ahead, marching to the beat of war drums.

As we stand on our high towers to watch this imposing battalion appear on the horizon, it is imperative for each one of us to formulate our plan.

What jobs will survive the AI-dominated future?

What skills do we need to develop?

How should we prepare our children and grandchildren?

Is the educational system sufficiently equipping our children for the future? If not, what should change? How can it be changed?

What should we do at home to help our children develop the necessary skills for the future?

Clearly, the more we know, the better we can prepare for this modern invasion—one that will change the way we live.

Ready or not, the vanguard of this mighty force is poised to march into man's domain.

Its marines have already crossed our borders.

The die has been cast!

Chapter 1
AI in Transportation and Logistics

Some of the most dramatic episodes in the history of man involved transportation. Astride a horse, a camel, or an elephant, man began to explore and expand his territory. Chariots became a symbol of political and military might. Carriages defined luxury and class in mobility.

Ships enabled voyages across the seas, leading to the discovery of new worlds. This mode of transportation then facilitated the establishment of colonies that were remotely controlled by affluent kingdoms and empires halfway around the world. Commercial vessels brought about intercontinental transport of tourists and immigrants, as well as international trade of general cargo such as cows, sheep, corn, tea, pulp, paper, steel, and much more. Today, highly advanced super Post-Panamax vessels can carry thousands of cargo containers that each weigh several metric tons.

Highly sophisticated military ships—cruisers, destroyers, frigates, aircraft carriers—now transport thousands of personnel as well as weapons, supplies, and equipment from one continent to another in a matter of days.

The invention of the automobile ushered in a new era of land transportation. It created the need for nationwide infrastructure such as roads and bridges that connected cities, towns, and remote rural areas. Commercial automobiles, also known as taxis or limousines, paved the way for accessible public transport. The car industry likewise created status symbols for a range of social classes—from the basic sedan of the salaried employee to the fancy sports cars and luxury vehicles that distinguished the rich and famous.

The development of more powerful engines made new forms of "mass" land transportation possible—buses, trucks, and trains—for moving large numbers of people and hauling heavy loads of goods.

Today, buses provide affordable public transportation for millions of commuters around the world, and have become so integral to the life of the general population who rely heavily on this mode of transportation to bring them to work and drive them back home.

Trucks move a wide range of cargo—food, office supplies, building equipment, military gear, construction materials, and many others—from state to state and across national borders.

Trains transport even more people and cargo—and do so with remarkable speed and efficiency. Some trains can now travel at top speeds of over 200 miles per hour.

Beyond all those, one of the most exceptional achievements in transport engineering is the airplane. We now have planes that carry hundreds of passengers from one state, country, or continent to another. We have jet planes that can breach the speed of sound and smoothly maneuver at supersonic speed. We even have spaceships that can reach the moon and other planets.

Yet throughout mankind's engineering history, the creation of autonomous, self-driving vehicles was not perhaps foreseen by many. Some may have dreamt of this possibility, but even world-famous sci-fi movies still showed futuristic cars steered by human drivers.

Whatever shortfall in imagination we may have had in the past, AI technologists are now making up for it.

Autonomous Cars

Today, Google is perfecting its Waymo fleet of self-driving cars to serve the public. Waymo cabs have been undergoing road tests for tens of thousands of hours.

It may not be too long before we will be able to request for Waymo taxis using a mobile app. Here is Waymo's mission statement:

> Waymo's mission is to make it safe and easy for people and things to move around. We aim to bring fully self-driving technology to the world that can improve mobility by giving people the freedom to get around, and save thousands of lives now lost to traffic crashes.[10]

Since 2016, a startup company based in San Diego, California called AutoX has been developing a product that you can install in your car to transform it into a self-driving vehicle. The company, which was founded by Professor X (Jianxiong Xiao) from MIT, aims for "the democratization of logistics," i.e., providing logistics solutions for small- to medium-size players like restaurants and grocery stores.[10] Just imagine a medium-size grocery store having a few self-driving cars that can fulfill online orders without having to rely on a big logistics company.

In October 2019, people on the streets of London saw for the first time some self-driving taxi cabs.[11] Below is an excerpt from an article that was released in *The Sunday Times* website on October 8, 2019:

> SELF-DRIVING taxis have hit the streets of London for the first time during a week-long trial in the capital.
> The culmination of a 30-month development process led by the government and industry-supported DRIVEN autonomous vehicle technology consortium, the tests saw a collection of

> Ford Mondeo-based test cars complete short runs on a pre-programmed course on public roads through Stratford, in the east of the city, a short distance away from the Queen Elizabeth Olympic Park, venue for the 2012 Games.
>
> While this isn't the first time autonomous vehicles have been tested in an urban environment (the same self-driving research vehicles were put through their preliminary paces in Oxford earlier in the year), DRIVEN said these tests have been "the most ambitious" yet, due to the demands that come with driving in a megacity.[12]

Just how serious are the automakers in making this sci-fi dream a reality?

On December 23, 2019, *Digital Trends* released an article titled "GM requests green light to ditch steering wheel in its self-driving cars." Below is an extract from that article:

> Cruise, the autonomous-car unit of General Motors, wants to start testing its self-driving cars sans steering wheel. Indeed, it's already produced a video showing what such a design might look like.
>
> Current National Highway Traffic Safety Administration (NHTSA) regulations, which were created at a time when self-driving cars were the stuff of science fiction, mean that vehicles have to include essential safety features such as a steering wheel and brake pedal. But the emergence of robot cars has put these rules under the spotlight.
>
> According to a recent Reuters report, the NHTSA has confirmed it's in talks with GM, which is asking for permission to test a limited number of autonomous cars without steering wheels or any other type of manual controls.[13]

Other countries outside the United States are equally determined to show superiority in this competition.

In a powerful demonstration of its resolve to win the AI race, the Chinese tech giant Baidu has secured licenses to test their self-driving cars in the Chinese capital.

Venture Beat released an article on December 31, 2019 announcing that "Baidu secures licenses to test self-driving cars in Beijing." Here is a portion of that herald:

> Tech giant Baidu is slowly but surely progressing toward the launch of a commercial robot-taxi fleet in mainland China. This week, the company announced that it has secured 40 licenses to test driverless cars carrying passengers on designated roads in Beijing, making it one of the first to do so in the Chinese capital. It also revealed that its self-driving cars have traveled more than three million kilometers (or about 1.8 million miles) during tests in 23 Chinese cities to date, up from 2 million kilometers (roughly 1.2 million miles) across 13 cities as of July.[14]

You may have already figured out the gist of the story that is about to unfold. Someday soon, your daily ride to the office will involve talking to your car from the backseat, saying, "Take me to 56th St. corner 5th Avenue, then find a parking garage." Your car will take you to your destination and will subsequently proceed to a parking garage. Of course, your car will electronically transact via some wireless protocol with the computer system of the parking garage. After you are done with work, you will text your car, which will electronically signal the parking garage system to allow your car to exit and charge your credit card. Your car will know where to pick you up as it can track the location of your mobile phone.

Obviously, this scenario will not happen overnight. Nonetheless, the business opportunities presented by auto-driving technology are promising.

On September 10, 2019, Goldman Sachs researchers predicted that the market for self-driving technology will grow to $96B by 2025 and to $290B by 2035.[15]

The stakes may be high, but the business potential cannot be ignored.

This prediction points to one important message: Self-driving vehicles are likely to become a reality in the not-so-distant future.

Autonomous Trucks and Buses

The Port of Rotterdam, a city in South Holland, The Netherlands, is the largest container shipping port in Europe.

Avid fans of shipbuilding history know that the Dutch are among the world leaders in that domain. In fact, back in the 1600s, the ship building industry was dominated by the Dutch, the English, and the French.[16]

Fast forward to 2015, the Dutch's pioneering spirit resurfaced on the global stage. This time, they launched their self-driving trucks inside the yard of the Port of Rotterdam—but not without labor concerns.[17] As a side note, you can easily search for videos on the internet showing driverless trucks carrying containers from one location to another inside the Port of Rotterdam.

Soon thereafter, Europe started testing self-driving trucks and buses on their public roads. Here is an excerpt from an article titled "Convoy of self-driving trucks completes first European cross-border trip" that was published by *The Guardian* on April 7, 2016:

> Six convoys of semi-automated "smart" trucks arrived in Rotterdam's harbour on Wednesday after an experiment its organisers say will revolutionise future road transport on Europe's busy highways.
>
> More than a dozen self-driving trucks made by six of Europe's largest manufacturers arrived in the port in so-called "truck

> platoons" around midday, said Eric Jonnaert, president of the umbrella body representing DAF, Daimler, Iveco, MAN, Scania and Volvo.
> "Truck platooning" involves two or three trucks that autonomously drive in convoy and are connected via wireless, with the leading truck determining route and speed.
> Wednesday's arrival concluded the first-ever cross-border experiment of its kind, with self-driving trucks leaving factories from as far away as Sweden and southern Germany, Jonnaert said.[18]

Today, major bus and truck manufacturing companies are prototyping and testing their next-generation vehicles with auto-driving features.

In 2018, Traton, a truck and transport services company, deployed one self-driving truck in a mining operation in Australia managed by Rio Tinto.[19] Here is an excerpt from Traton's website:

> Scania is testing a new generation autonomous transport system at Rio Tinto's Dampier Salt operations in Western Australia. The first phase of the trial started in August 2018 and involves a Scania XT 8×4 autonomous tipper truck working separately from Dampier's active operations.
> During this initial stage, a safety driver rides in the vehicle to observe the truck's performance and, if necessary, intervenes. In subsequent phases, additional autonomous Scania trucks will be added to develop vehicle-vehicle awareness and intelligent fleet supervisory controls.
> Rio Tinto's head of Productivity & Technical Support, Rob Atkinson says, "We're pleased to be trialing this technology in trucks that are smaller than our traditional haul trucks. This has the potential to give us more flexibility in the way we operate in a number of areas across Rio Tinto. We've seen automation create safer and more efficient operations in our business and

> this is a next step in evaluating options for delivering further improvements through the use of technology."[20]

In the summer of 2019, countries such as China, Singapore, the UK, France, and Germany started testing self-driving buses on their streets. The country with the least restrictions will likely win this race into the next era of transport technology.

But beyond driverless, something else is hovering on the horizon.

Flying Cars

Reality has a way of imitating fiction. When something previously believed as fantastical all of a sudden becomes real, we hear people say, "Reality is stranger than fiction."

Such is the case with flying cars.

Over and above the remarkable developments in the transportation industry, designers and engineers are now building and testing flying cars using drone technology.

Imaginations are soaring and the big players are coming to place their bets.

On May 9, 2018, *NBC News* released an article titled "Uber just unveiled a prototype of its futuristic air taxi." The news bit pointed out that the Uber Air taxi will be able to vertically take-off and land and will travel at speeds of up to 200 miles per hour.[21]

On July 17, 2018, *USA Today* ran a story on the Volante Vision Concept, a highly sophisticated design idea of flying cars from the British luxury automaker, Aston Martin. The article titled "Aston Martin's flying car concept is fit for James Bond" cited vertical take-off and landing capabilities as well as self-driving features. According to the article, the Aston Martin flying car, which is designed to carry three passengers, will not be available in the market until around 2025.[22]

In late 2019, Singapore announced the construction of a pad for flying cabs. A *Straits Times* article that was published on October 22, 2019 bore the headline, "Flying taxis could soon take off in Singapore."[23]

Here is an excerpt from that article:

> Some people may joke that we are not yet in the future because there are no flying cars.
> But, within five years, taking a flying taxi between Marina Bay and Changi Airport could become a regular occurrence.
> German aviation firm Volocopter held its first public test flight along the Marina Bay waterfront on Tuesday (Oct 22).

While the Germans, the British, and the Americans are busy working on their flying cars, Chinese technologists are determined to be the first in the world to field these for commercial operation.

On November 1, 2019, *The Verge*, an online magazine that focuses on technology-related news, released an article about a Chinese drone company called Ehang. The article, titled "Autonomous flying taxi startup Ehang files $100 million initial public offering," stated that Ehang is going public on Nasdaq to raise $100M. Their business objective? To "launch what it hopes is the world's first autonomous air taxi service in the coming months."[24]

With all the haste going on, it is possible that we will soon see flying cars zipping along our city skylines.

However, despite all the excitement over flying cars, there are also numerous challenges to overcome. The first hurdle is the acceptance of the general public, especially after the shocking helicopter crash that took the life of American basketball superstar, Kobe Bryant.[25]

Secondly, there will have to be laws and regulations around the use of flying cars. Drivers of such vehicles will have to be properly trained and licensed to ensure the safety of every passenger as well as of the general public. Insurance will undoubtedly be an important part of that conversation.

Photo by João Rocha courtesy of Unsplash

Will drone taxis be one day driverless as well? It is not a remote possibility. Manufacturers of flying cars are already integrating self-driving capabilities in their conceptual designs. The foundational technologies like Global Positioning System, machine learning, and machine perception are already available. It is now a matter of putting them together and incorporating them in drones.

Drones

Contrary to what many believe, drones or unmanned aerial vehicles (UAVs) are not exactly a new invention. During World War I, the first-generation drones were pilotless airplanes that were radio-controlled.

The first successful UAV developed by the US military around 1918 was called *Kettering Bug*. This was a self-flying torpedo developed under the supervision of Charles F. Kettering and Orville Wright, one of the Wright brothers who pioneered American aviation.[26]

Today, UAVs come in different shapes and forms. In fact, drones have flown their way into various commercial applications. They are now being used to secure geographic borders, apprehend criminals, and assist in search and rescue operations. They film aerial views and commercial footage for marketing golf courses, hotels, and vacation resorts. Drones are also used for inspecting remote areas, surveying land, monitoring the environment and climate, viewing crop production and crop growth, and even keeping tabs on livestock.

It is likewise worthy to note that major logistics companies have been testing drones to deliver goods to their customers' doorsteps.

Among them, the giant logistics company, Amazon, has been working on a delivery system called *Prime Air*[27] using self-driving aerial vehicles or drones.

Even the global conglomerate Unilever sees big business opportunities using drone technology. On February 6, 2020, *Futurism* (futurism.com) released an article titled "BEN & JERRY'S IS TESTING A DRONE DELIVERY SYSTEM FOR ICE CREAM." Below is an excerpt:

> On Tuesday, drone startup Terra Drone Europe announced that it had partnered with Unilever, the parent company of Ben & Jerry's, to test a drone delivery system for the ice cream maker's beloved ice cream cups.[28]

Ice cream cups today, can be many different things tomorrow. Have drone, will deliver!

It is not hard to imagine that in the coming years drones will be equipped with artificial intelligence that will enable them to autopilot, avoid air collisions, find alternate aerial routes, and even record receipt of deliveries on video without human intervention.

The future of logistics will then involve a highly advanced network of drones that can deliver toiletries, snacks, sandwiches, pizzas, shirts, lipsticks, and books to wherever you may be.

Within minutes.

Top quality personalized service.

Any company that cannot provide such expeditious delivery will likely lose the competition.

On the other hand, this can also mean that, given a fleet of drones, a small company or restaurant can compete with the giants.

While technology can generally be a competitive advantage for large companies to dominate the market, clearly technology also has a way of leveling the playing field.

The Big Picture

Considering these numerous potential applications of AI in transportation and logistics, it is not difficult to extrapolate these developments to form a picture of what the future will look like.

One can conclude that there will be a progressive decline in job opportunities for human drivers—taxi drivers, bus drivers, and truck drivers—due to advancements in self-driving technology.

Reputable institutions such as Oxford Economics, Bloomberg, and Goldman Sachs have predicted as much regarding the potential impact of AI and automation on the labor market.

On the flip side, there will be new job opportunities that will be created by the same AI and automation technologies, including a variety of technologies related to self-driving vehicles like automated parking systems for autonomous vehicles, among many others. Additional jobs will be created along the lines of marketing, design, manufacturing, and testing of autonomous vehicles, maintenance and repair of self-driving vehicles, and more.

The bigger question, however, is this: Will these new opportunities make up for the jobs that will be lost due to autonomous vehicles?

The use of self-driving technologies will likely increase the profit margins of transportation and logistics companies. The rich will become richer. The poor will become poorer. As the few wealthy will become wealthier with the use of automation, the burden of supporting the jobless will be on the shoulders of the government in countries where such technology is emerging.

How will governments support millions of jobless citizens?

Will there be a need for the national leadership to provide a universal subsidy or what others call *Universal Basic Income* (UBI)?

Where will the money come from?

Is UBI a foundational principle of a capitalist country with a free market economy?

A massive re-training program is clearly the solution. The question then becomes, "Train the people on what?"

What new skills can a senior person, who has been driving taxi cabs for 30 years, acquire?

Will drivers who earn measly daily incomes have the time and the spare money to learn new skills?

Will there be enough jobs requiring those new skills?

How long will it take for a laborer to learn a completely new technical skill?

What are the chances of senior people competing against younger people in the job market?

Some people have said that AI can either be the best thing or the worst thing that can happen to mankind.

Can it be that AI and automation will usher in a new era where people will do less work, receive universal basic income, and have more time to pursue what they really want to do? Will that not be nirvana or heaven on earth?

On the other hand, if millions of people will lose their jobs and be unable to feed themselves and their families, leading to a rise in crime—will that not be the hell we all dread?

The plot can change its course, however, depending on how the key institutions of each nation will address this looming technological trend.

Indeed, AI, like time, will move forward whether we like it or not. In the final analysis, it will be absolutely imprudent to ignore the signs.

The future is taking shape and we will see its true form much sooner than we think.

The message is clear.

We have to prepare.

Now.

Chapter 2

AI in Healthcare

"AI winter" is a term used to describe the years when investments in artificial intelligence dipped. The first AI winter happened in the early 1970s—about a decade after the term "artificial intelligence" was coined by pioneers of AI.

The second AI winter occurred around the late 1980s to the early 1990s when, once again, funding for AI projects dropped.

How does the concept of AI winter fit in this chapter on healthcare?

If AI investments in healthcare today are an indication, then one can conclude that the most recent AI winter is over.

A reputable data mining and analytics company called CB Insights released an article on July 9, 2019 titled "Artificial Intelligence Deals Tracker: ~7K Deals Across 20 Industries in One Heatmap." The article highlighted the following:

- From 2013 to the present, more than 3,600 AI startups worldwide have raised equity funding.
- About $66B has been invested in AI startups from the second quarter of 2013 to the second quarter of 2019.
- Healthcare is the top favorite in AI investments.

- AI startups in healthcare are using AI for diagnostics, drug R&D, remote patient monitoring, emergency room and hospital management, genomics, and patient risk scoring, among other healthcare applications.[29]

Let's take a look at some examples of how AI is being applied in the healthcare industry today.

AI in Medical Diagnostics

In the area of medical diagnostics, AI is proving to be valuable in pathology, radiology, and even endocrinology.

Pathology is a branch of medicine that deals with the laboratory examination of body tissue samples—also generally known as biopsy.

Artificial intelligence is now being incorporated in an approach in diagnosis called Computational Pathology. The methodology is straightforward yet, nonetheless, impressive. AI is fed with digital images of specimens and biopsies. Each image is tagged with a relevant medical diagnosis: cancer, benign, normal, etc. As a result, when AI is presented with a new digital image of a biopsy, it is able to provide a medical diagnosis based on a vast knowledge base of medical images it previously learned.

Companies like Pixel Scientia Labs,[30] Proscia,[31] and Paige.AI[32] are racing to build computing ecosystems for digital pathology that will aid to accelerate biomarker discovery and also help generate new insights for human pathologists.

Below is an excerpt from an article titled "Startups Bring AI to Pathology" published by the American Association for Cancer Research (AACR) on February 23, 2018:

> An early pioneer of computational pathology, Thomas Fuchs, PhD, of Memorial Sloan Kettering Cancer Center (MSKCC) and

> Weill Cornell Medicine, both in New York, NY, announced the launch of a new academic spinoff company this month. The startup, known as Paige. AI aims to replace pathology's glass slides and microscopes with digitized images and artificial intelligence (AI).
>
> It's hardly alone in this endeavor. Software giants, medical device manufacturers, and dozens of small startups are developing pattern-recognition algorithms to help pathologists more accurately spot tumors on digitized images of tissue. However, with $25 million from investors, intellectual property from MSKCC, and exclusive rights to the cancer center's library of 25 million pathology slides—one of the world's largest tumor pathology archives—Paige.AI has established itself as a frontrunner in the development of a clinical-grade AI tool to guide cancer diagnosis and treatment.
>
> "We have data at a scale no one else has," says Fuchs, the company's founder and chief scientific officer. Paige.AI is first deploying its machine-learning models, image classification software, and high-performance computers to diagnose prostate cancer. Algorithms for breast cancer, lung cancer, and other tumor types are also in development.[33]

Imagine, AI being fed with hundreds of thousands of medical images—much like a seasoned pathologist undergoing many decades of highly specialized training.

The only difference between a human pathologist and an AI system is that the latter does not forget and does not get tired. AI's "eyes," unlike those of a human doctor, will not progressively fail as years go by.

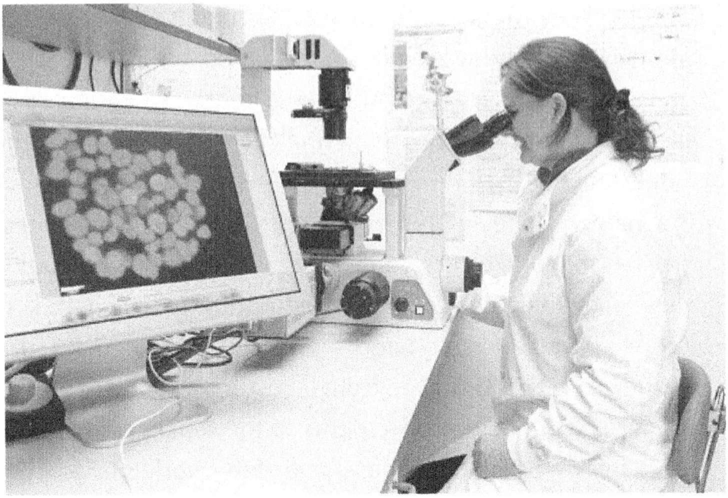

Photo of digitized specimen

Before we all get perturbed about the possible risks, the upside is that this AI technology will enable pathologists to discover biomarkers with greater speed and accuracy. Clearly, a human pathologist will still have to consider other corroborating evidence leading to a diagnosis.

In addition, because the pathologic medical images are digitized, they can be transmitted over the internet and analyzed by an AI-enabled machine in another location. This means that doctors in remote and less privileged areas can have access to such AI-enabled platforms to aid them in diagnosing their patients' conditions.

Artificial intelligence is also being applied to Digital Radiology Imaging. Similar to the process for Computational Pathology, AI is fed with hundreds of thousands of digital radiologic images. Each image is tagged with a diagnosis: tuberculosis, pneumonia, normal, etc. Thus, when AI is presented with a new digital radiologic image, AI will be able to suggest a diagnosis.

Again, imagine the increased efficiency this can provide to radiologists. The potential business value to many hospitals and clinics around the world cannot be ignored.

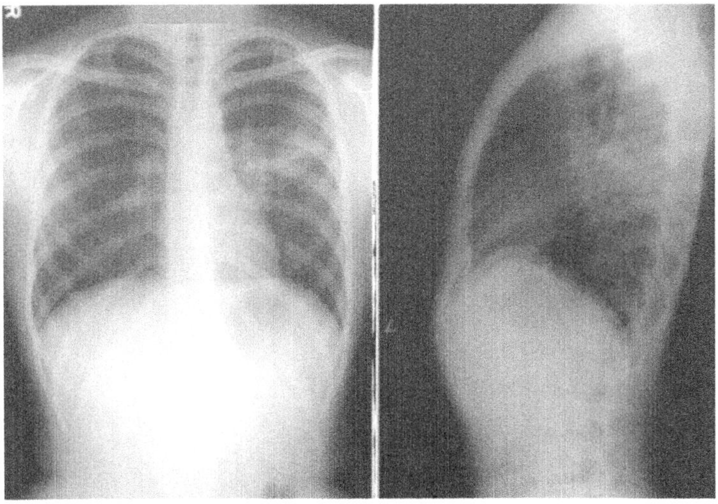

Photo by CDC courtesy of Unsplash

According to a February 4, 2019 article in *Health Imaging* (healthImaging.com), there are over 120 startups that are developing AI-enabled solutions in the area of digital imaging. Here is a section of that report:

> In 2018, capital investments in startup companies developing medical imaging AI solutions reached almost $580 million—more than double the 2017 amount of $270 million, according to a new market report published Jan. 31 by Signify Research.
> Of the 2018 total, HeartFlow—the Silicon Valley startup founded in 2007, known for its noninvasive AI analysis software that creates 3D models of coronary arteries from CT scans—accounted for $240 million.
> Currently, more than 120 startups are developing AI solutions for medical imaging, according to the report. Since 2014, these companies have accumulated more than $1.2 billion in investments.
> Of the 10 most-funded companies in 2018, five were from the U.S., four were from Asia and one was European. However,

> 2018 saw 10 new AI companies enter the market, a significant decline since its 2016 peak of 26 companies.[34]

It goes without saying that funding is a major factor in determining whether a technology will become a reality or not. With capital investments at over half a billion dollars, one can say with some degree of confidence that AI-enabled radiologic imaging systems will one day become a norm.

The value is undeniable.

Investors have joined the party.

The bets are on the table.

The dice are now rolling.

AI in Surgery

Surgery is another area in healthcare where intelligent machines will play a compelling role moving forward.

Robotic surgery, or robot-assisted surgery (RAS), is a relatively new application of robotics in this field. It is also often referred to as *minimally invasive surgery.*

RAS platforms today enable surgeons to perform procedures that were deemed too delicate, difficult, or impossible two decades ago. With RAS platforms, the issue of a surgeon's hand tremor is eliminated. As stated above, robotic surgeries are minimally invasive, i.e., small incisions, resulting in less blood loss, reduced pain, and faster recovery.

How does robot-assisted surgery work?

A surgeon sits comfortably on a platform from which he controls the robotic arms to perform the surgery. The robotic system makes a small incision in the patient's body through which miniaturized instruments and a highly advanced high-definition camera are inserted to probe the target organ and perform the surgical operation. Typically, RAS platforms come with large high-definition

TV monitors, similar to those in video game systems, that allow the surgeon to clearly view every stage of the operation.

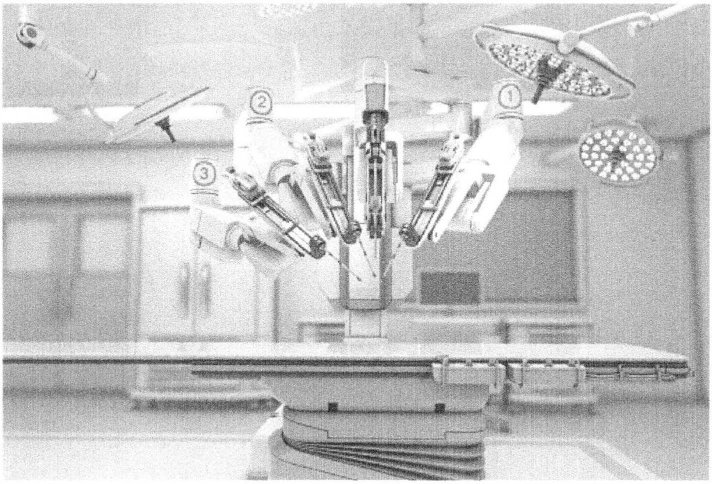

Photo of robot assisted surgery platform

Probably the most famous RAS platform is the da Vinci surgical platform developed by Intuitive Incorporated.[35]

Five other platforms were mentioned in a *Robotics Business Review* (RBR) article that appeared on RBR's website on December 31, 2018: The PRECEYES Surgical System, The CorPath System, The Monarch Platform, The Mako Rio, and The Versius.[36]

It is easy to appreciate why minimally invasive surgery is good for patients, but what are the benefits of RAS technology to surgeons?

First of all, RAS platforms can make some extremely challenging surgeries a bit more manageable for surgeons. It is generally believed or accepted that RAS is good for colon and gynecological surgeries because of its ability to maneuver within tiny spaces.

More importantly, RAS is kinder to a surgeon's body. Surgical operation is a stressful activity and is broadly harsh on a surgeon's eyes, neck, shoulders, arms, and feet. RAS platforms help ease the strain for surgeons as they can sit and control the RAS platform with minimal stress to their bodies.

After all, let us not forget that surgeons are humans too. They get tired after prolonged effort just like any normal human being. That said, any patient who is scheduled to undergo a surgical operation would not wish his or her surgeon to be tired and stressed while performing the procedure. Therefore, a RAS platform which allows the surgeon to sit comfortably while controlling the robotic system is a welcome development.

While more and more hospitals are recognizing the value of RAS platforms, the current downside to this technology is its multimillion-dollar price tag, plus costly annual maintenance.

However, as competition in the RAS industry becomes stiffer, RAS platforms are expected to become more affordable. When that happens, we can expect robot-assisted surgeries to become a regular item in hospital service catalogs. We can also expect RAS platforms to become less bulky, more intelligent, and much more ergonomic in the coming decades.

Ultimately, as RAS platforms become economically feasible for more hospitals, all three parties—hospital, surgeon, and patient—will clearly benefit from the investments.

AI in Drug Research

It is widely known that drugs such as penicillin and aspirin have saved and continue to save the lives of millions of people. As a result, life expectancy has significantly improved over the last century.

However, the pursuit of new cures is a lengthy process and often comes with a hefty price tag.

Unbeknownst to many, it takes over $2.5 billion to develop a new drug.[37] This estimate is expected to grow over time due to increasing costs of researchers, equipment and materials, and numerous other related fees.

The other hurdle in the development of a new drug is that it takes approximately 10 to 15 years to complete three phases of clinical trials.[38] A vital precaution to ensure that every medicine is safe for public consumption.

That being said, unfortunately, not all drugs pass the required clinical trials. Every drug research that fails to secure the green light translates to loss of revenue for a pharmaceutical company. This lost revenue will either be written off or will need to be somehow recovered from other revenue sources. The key takeaway is that research costs are ultimately passed on to consumers.

In that case, is there a better and more expeditious way to discover new medicines?

A thrilling development in the field of drug research is the application of artificial intelligence in the drug discovery process.

With AI, the timeframe for drug discovery can be greatly reduced. This will then hopefully result in enhanced speed to market and more affordable medicines for consumers.

On May 25, 2018, the *Pharmaceutical Technology* website released an article titled, "Is artificial intelligence the future of drug discovery?" The following excerpt reveals the exciting developments:

> Increasing investments in AI in drug discovery by big pharma suggest a truth behind the benefits of applying machine learning to identify and screen potential drug candidates. More and more, big pharma is partnering with AI-driven companies in hopes of more accurately predicting drug candidates and cutting R&D costs and time, prompting GlobalData to ask—Is AI the future of drug discovery?
>
> Pharmaceutical giant Merck is one such company taking a lead in implementing AI-based solutions in drug discovery. Merck entered the AI space early, in 2012, striking a partnership with Numerate, an AI-based company leveraging algorithms and cloud computing to transform the drug design process. The collaboration was initially developed for Merck to utilize

> Numerate's computer-based drug design technology to develop novel small molecule drug leads for an undisclosed cardiovascular disease target.
>
> In addition, Merck is working with Atomwise, the creator of AtomNet, which uses deep learning technology for the discovery of novel small molecules. Although the project is confidential, Merck is leveraging Atomwise's AI-based technology to scan existing medicines that could be redesigned to fight old and upcoming diseases.
>
> Merck is just one of many pharmaceutical companies partnering with AI-focused companies to advance drug discovery. Celgene partnered with GNS Healthcare to utilize its Reverse Engineering and Forward Simulation causal machine learning and simulation platform; GSK entered a $43M drug discovery collaboration with UK-based AI-driven startup Exscientia; Pfizer entered collaboration with IBM Watson for immuno-oncology drug discovery research, and the list goes on.[39]

What we can hope to come out of this collaboration between AI companies and pharmaceutical companies is acceleration in the discovery of much-needed cures for diabetes, various types of cancer, cardiovascular diseases, and many more.

If drug companies succeed in lowering their research costs, then that could also mean more affordable medicines for the general public.

The good news is it is not only the drug companies that are unveiling new cures.

On February 20, 2020, *MIT News* released an article titled "Artificial Intelligence yields new antibiotic." The article announced that MIT researchers have found a new antibiotic compound through the use of a machine learning algorithm. What's more, the drug demonstrated the ability to kill a number of the world's most deadly bacteria as well as some strains that are known to be antibiotic-resistant.[40]

Due to the exciting potential of AI-enabled drug research, many believe that life expectancy in the coming decades may reach up to 100 years and above.

In the event that life expectancy reaches 120 years, will that necessarily mean that retirement age will go up to 80 or 90 years old?

If, or when, that happens, the obvious challenges will involve food and water, shelter, eldercare, social security pensions, Medicare, insurance, among others.

Broadly speaking, will it be truly beneficial to societies if people do live up to 120 years of age?

Will the Earth's resources be able to sustain such prolonged human life expectancy?

Indeed, the promise of a longer life is not going to be without its problems.

Such is life.

Or in French, *"C'est la vie!"*

Personalized and Interactive Healthcare

Many believe that the new generation is more health conscious than prior generations. Case in point, there are more gyms now than, say, 50 or 60 years ago.

Of course, one can argue that several decades ago people were much more physically active than the general population today. Back then, people walked miles to buy groceries, planted vegetables in their backyard, played sports with neighbors, biked around the community, did their own woodwork, physically washed their clothes and dried them out in the sun, just to name a few.

Regardless of what one believes, investors are now betting on technologies that aim to capitalize on the health consciousness of the new generation. There are a number of companies, both startup and established ones, that are working towards delivering what is called personalized and interactive healthcare solutions.

What this means is that a software application will be like your personal medical assistant. The system will have enough information about you to remind you when you should take your medicines. You can even interact with the application in the same way that you converse with a human doctor. The software will ask you about signs and symptoms that you are experiencing and then suggest a diagnosis based on your answers.

As you can already guess, this kind of software, which has your health information stored somewhere on the internet, can be quite controversial to many.

Nonetheless, a good example of such a chatbot software is one called Babylon Health.[41]

On August 2, 2019, *TechCrunch* released an article on Babylon Health's successful fundraising. Key details are in this excerpt:

> Babylon Health, the U.K.-based startup that has developed a number of AI-based health services, including a chatbot used by the U.K'.s National Health Service to help diagnose ailments, has confirmed a massive investment that it plans to use to expand its business to the U.S. and Asia, and expand its R&D to diagnose more serious, chronic conditions. It has closed a $550 million round of funding, valuing Babylon Health at more than $2 billion, it announced today.
> This is the largest-ever fundraise in Europe or U.S. for digital health delivery, Babylon said.[42]

There are numerous videos on the internet that demonstrate the power of Babylon Health's chatbot. While there is obviously no comparison to a doctor physically examining you, Babylon Health can be accessed anytime, anywhere. Where there is access to the internet, there is access to Babylon Health.

If a $550M investment is any measure of investor confidence in what this AI-enabled technology can deliver, then there is clearly tremendous hope in Babylon Health's potential.

However, Babylon Health is not alone in this arena. There are a number of aspiring competitors like HealthTap,[43] Sensely,[44] Push Doctor,[45] Your.MD,[46] Doctor Care Anywhere,[47] among others. Even the giant tech company Google[48] has joined the competition.

Now that doesn't mean though that the bandwagon is not facing some storms.

On September 18, 2019, *BBC News* released an article titled "Google swallows DeepMind Health." Below is an excerpt:

> Last November the pioneering London AI firm DeepMind announced that its health division was being transferred to its parent company Google.
>
> Now, 10 months later, the move is finally happening, the delay a measure of how controversial and painful this process has been.
>
> When the restructuring was announced last year, DeepMind was accused of breaking a promise not to hand over any NHS data to the American search giant.
>
> Google said the transfer to its global health division would give DeepMind's Streams app, which helps doctors and nurses to monitor patients with a severe kidney condition, the resources it needed to expand internationally.
>
> The Streams app had caused quite a stir when it was revealed that it used the data of about 1.6 million patients, without asking permission. The Information Commissioner's Office (ICO) ruled that the hospital involved in the research—the Royal Free in London—had not done enough to protect patient privacy.[49]

As you can deduce from the article, there is controversy around companies like Google having access to people's health information without their express permission.

However, if this challenge can be hurdled, then Google DeepMind Health, together with other major players in this league, will undoubtedly change the way we view healthcare in the future.

Mind-controlled Prosthetics

In its simplest form, AI means intelligent machines—machines that can think and act like human beings.

What about a machine that behaves like a part of a human body? That type of machine has been in existence for decades. It is called a prosthetic.

Photo of prosthetic arms

The field of prosthetics has helped thousands, if not millions, of people around the world who have lost their arms or legs due to disease, an accident, or injury in military service.

The good news is that some form of artificial intelligence is now being incorporated into prosthetics—like the recently developed mind-controlled robotic arms.

The first-generation mind-controlled robotic arms required brain implants to detect neural signals. However, on June 19, 2019, Bloomberg[50] released an article titled "First-ever successful mind-controlled robotic arm without brain implants." This excerpt shares the breakthrough news:

> A team of researchers from Carnegie Mellon University, in collaboration with the University of Minnesota, has made a breakthrough in the field of noninvasive robotic device control. Using a noninvasive brain-computer interface (BCI), researchers have developed the first-ever successful mind-controlled robotic arm exhibiting the ability to continuously track and follow a computer cursor.
>
> Being able to noninvasively control robotic devices using only thoughts will have broad applications, in particular benefiting the lives of paralyzed patients and those with movement disorders.
>
> ...
>
> Bin He, Trustee Professor and Department Head of Biomedical Engineering at Carnegie Mellon University, is achieving that goal, one key discovery at a time.
>
> "There have been major advances in mind controlled robotic devices using brain implants. It's excellent science," says He. "But noninvasive is the ultimate goal. Advances in neural decoding and the practical utility of noninvasive robotic arm control will have major implications on the eventual development of noninvasive neurorobotics."
>
> Using novel sensing and machine learning techniques, He and his lab have been able to access signals deep within the brain, achieving a high resolution of control over a robotic arm. With noninvasive neuroimaging and a novel continuous pursuit

> paradigm, He is overcoming the noisy EEG signals leading to significantly improve EEG-based neural decoding, and facilitating real-time continuous 2D robotic device control.
> Using a noninvasive BCI to control a robotic arm that's tracking a cursor on a computer screen, for the first time ever, He has shown in human subjects that a robotic arm can now follow the cursor continuously. Whereas robotic arms controlled by humans noninvasively had previously followed a moving cursor in jerky, discrete motions—as though the robotic arm was trying to "catch up" to the brain's commands—now, the arm follows the cursor in a smooth, continuous path.[51]

The obvious challenge with mind-controlled prosthetics is affordability. When people who need the prosthetics can't afford to buy them, then the ultimate objective, however noble it may be, is not met.

Thankfully, there is light at the end of the tunnel.

A company called Open Bionics[52] has been working on what seems to be a plausible solution: 3D printed bionic limbs. Their biggest win? Children love their designs. Why? Their bionic arms look like superhero costumes!

On June 28, 2018, the *Access and Mobility Professional*[53] website published an article titled "Prosthetics firm targets NHS with affordable 3D-printed super hero arms." The encouraging account is excerpted here:

> Prosthetic and bionic technology has come a long way in recent years as more advanced components and functions are squeezed into products, allowing users to perform more complex actions. Some researchers have even begun developing prototype products that allow the user to feel as though they are using a biological limb.
> But some barriers have remained constant to manufacturers even as technology has advanced. Creating equipment that is

low-cost to produce is difficult in prosthetics and making items that users, and particularly children, want to wear and are not embarrassed by is another huge challenge for suppliers.

Looking to tear down these barriers is a Bristol start-up called Open Bionics which is developing 3D printed bionic limbs that are cheap to produce and, importantly, exciting to wear for the user. It is targeting a gap in the market and if it takes off it could change the lives of users and establish the business as a leading player in the sector.

Samantha Payne, Open Bionics' co-founder, told an audience at Naidex recently that the company is trying to squeeze into the gap where competitor companies have been unable to fill because of their generally expensive technology. The high-cost of mainstream products currently on the market, she says, means few consumers are able to buy the products and the NHS is likewise unable to afford them.

Open Bionics uses 3D printing instead of metal to create bionic hands and limbs, which makes them cheaper and gives them potential to sell in greater volume to private customers as well as the NHS. In fact, it believes it can produce a personalised bionic hand for under £5,000 and much quicker than private providers, which have a price point of around £60,000 for a multi-grip bionic arm. The outfit's recently-launched Hero Arm claims to be the most affordable multi-grip bionic arm ever, at half the price of its nearest competitor.[54]

What kind of artificial intelligence is built in the bionic arms produced by Open Bionics?

Here is the secret sauce: There are sensors that detect signals from the residual limb of the amputee and convert them to the appropriate bionic hand movement. What an amazing and life-changing technology!

There is no doubt that these bionic parts will become more and more sophisticated in the coming years.

By becoming more affordable as well, there is no debate that this kind of intelligent machine can bring so much hope to millions of people in this broken world.

Intelligent Medical Devices

The field of prosthetics is not the only fertile ground for the application of AI. There has recently been some fascinating progress in the development of intelligent medical devices.

Among many, there are three devices I would like to mention: a self-diagnosing stethoscope, smart asthma monitoring, and an intelligent insulin pump.

Self-diagnosing Stethoscope

In a wealthy country where there is an abundance of healthcare professionals and sophisticated equipment, diagnosing pneumonia or a heart problem may be a straightforward task.

What about those in less fortunate locations?

Is it possible that some intelligent device could be developed to help medical workers in remote places diagnose pneumonia or other ailments?

It is not just possible. It has already happened.

The encouraging news is that the Johns Hopkins Applied Physics Laboratory [55] has developed an intelligent stethoscope that can diagnose pneumonia—even without the presence of a physician!

On February 7, 2019, *Futurism* (futurism.com) published an article with the headline "SMART STETHOSCOPE" CAN DIAGNOSE PNEUMONIA ON ITS OWN. Below is an excerpt from that news bit:

> The Johns Hopkins team is currently field testing prototypes of its smart stethoscope in Peru, Bangladesh, Malawi, and the U.S., and it has high hopes for the future of the final version of the device, which it expects to be "significantly cheaper" than the $500 electronic stethoscopes currently on the market.
>
> "By enabling on-the-spot screenings with a cheap device that can be handled by local health workers, we hope our reengineered scope will make a difference in the global health crisis of childhood pneumonia," the researchers wrote.
>
> "Viewed on the macro level, we hope to scale up our efforts so that hundreds of thousands of lives can be saved," they continued. "But on the micro level, we feel that if even one parent sees her child recover because of early diagnosis, all our hard work will have been worth it."[56]

How important is this development?

Given access to intelligent devices that can aid in diagnosing diseases like pneumonia, asthma, or even heart diseases, medical workers in far-flung areas can more efficiently diagnose patients and send them to the nearest hospital for treatment.

As they say, there are so-called reel heroes and there are real heroes. The Johns Hopkins Applied Physics Laboratory team definitely embodies the latter.

Smart Asthma Monitoring

Asthma is an illness that affects the quality of life of millions of people—both young and old—around the world.

Asthma is usually associated with various degrees of difficulty in breathing. Asthma attacks can be triggered by allergens like dust or pollen, or even by stress. Strenuous physical activity is also believed to potentially cause asthma attacks. Some attacks can be so severe that they result in death.

Fast-acting asthma medicines, called bronchodilators, are designed to address an imminent or ongoing asthma attack by causing the muscles in the person's lungs to relax and at the same time dilate the individual's bronchial airways.

Other asthma medicines are meant to prevent the onset of asthma attacks through a daily dose of inhaled steroids.

As you can imagine, difficulty in breathing is extremely uncomfortable at the very least. It is vital, then, that a person with a history of asthma is properly equipped to detect an imminent asthma attack. This is where an intelligent device can help.

How does smart asthma monitoring work?

A company called Health Care Originals[57] offers a patch monitor (with a rechargeable battery) that one can wear anywhere on the upper torso, front or back. That electronic patch monitors your breathing and sends information to a mobile app installed on your cellphone. Based on monitoring data, the app can detect an imminent asthma attack and advise you.

How?

By causing the electronic patch to vibrate to notify you. In case you have designated an alternate alert recipient, that person will also get a text notification when you are experiencing an attack.

With so many people of all ages suffering from asthma worldwide, this intelligent device can significantly improve the quality of life of millions.

Intelligent Insulin Pump

Diabetes is a metabolic disorder that affects the way the body processes carbohydrates. This results in elevated levels of glucose in the person's blood and urine.

Diabetes can cause many other serious health issues such as cardiovascular diseases, kidney malfunction, and loss of vision. It also causes poor blood circulation resulting in non-healing wounds that often lead to amputation of gangrenous extremities or limbs.

There are two types of diabetes.

The "milder" type is diabetes type 2, which is generally related to an individual's lifestyle, eating habits, as well as hereditary factors. This type of diabetes manifests itself anywhere from adolescence to senior years, depending on the factors mentioned above. Its rise in recent decades is associated with the increase in obesity among the general population.

The more severe form is type 1 diabetes, which is called juvenile diabetes or insulin-dependent diabetes due to the fact that it manifests at a very early age and makes the person dependent on insulin injections for the rest of his or her life.

Why?

Because type 1 diabetes is a condition that destroys the pancreatic cells that are responsible for producing insulin. Insulin helps regulate glucose in the blood stream. This means that, without regular doses of insulin, a person with type 1 diabetes can go into a diabetic coma and die. For a type 1 diabetic, then, insulin injections are not an option, but a necessity.

It is often said that necessity is the mother of invention.

Such is the case with the co-founder of Bigfoot Biomedical,[58] Bryan Mazlish, whose five-year-old son was diagnosed with type 1 diabetes in 2011. Some three years after, Bryan modified his own insulin delivery system (made up of a glucose monitor and an insulin pump), wrote a predictive algorithm that controls the insulin dosage, and developed an app that displays the results. In late 2014, Bryan co-founded Bigfoot Biomedical.

On January 13, 2020, Bigfoot Biomedical released an article on their website titled, "BIGFOOT BIOMEDICAL® ANNOUNCES INITIAL $45 MILLION OF A SERIES C FINANCING LED BY ABBOTT." Below is an excerpt from that website post:

> Bigfoot Biomedical, Inc. announced today that the company has raised $45 million in the initial tranche of a Series C equity financing. The round is led by Abbott with support from existing investors, including Quadrant Capital Advisors, Senvest Capital, Janus Henderson, and Cormorant Asset Management. Proceeds from the financing will support the completion of product development and FDA clearance for the Bigfoot Unity™ Diabetes Management Program, an injection-based digitized insulin dosing platform utilizing a proprietary, connected insulin pen-based system that integrates Abbott's FreeStyle Libre glucose sensing technology. Submission of Bigfoot Unity is anticipated in 2020 with a target launch by year-end 2020.[59]

I am certain that a lot of people can relate to the struggles of Bryan Mazlish. I myself have relatives and close friends who suffer from diabetes. It is my hope that someday soon their lives will be much more comfortable.

The Big Picture

As discussed in the introductory section of this book, one of the most basic questions we need to answer with regard to AI is this: Should we pursue AI, or should we abandon it?

While there is reason to be concerned about the potential of AI to displace a large percentage of the labor force, the application of AI in healthcare, on the other hand, can prove to be beneficial to millions of people worldwide.

The collaboration between AI companies and drug research companies can lead to expeditious discovery of cures for some of the most debilitating diseases known to man.

The development of intelligent prosthetics will give mobility and independence to those who have lost their limbs.

Intelligent medical devices such as the self-diagnosing stethoscope can be of great help to health workers in remote areas.

Robot-assisted surgery now makes it possible for surgeons to perform highly delicate operations that were deemed impossible before.

In effect, it can be argued that the benefits that may be derived from the application of AI in healthcare will far outweigh the negative impact of AI on the job market.

This is not to diminish the fact that the loss of millions of jobs will also have an adverse effect on the general health and well-being of many people. However, jobs can be created; in fact, robotics and AI will themselves give rise to new job opportunities. On the other hand, people will only be able to work for as long as they are strong, healthy, and not stricken by some crippling disease.

Therefore, the application of AI in healthcare cannot be abandoned.

You see, AI is like the wind that can be a storm to many, but a breath of fresh air to others.

The reality is that countries, companies, and individuals will continue to pursue AI research and development for various reasons—national security, military defense, profit, noble causes, or as a pure intellectual exercise.

The truth is that the impact of AI on you and me is left for us individuals to figure out.

The key question is this: As the winds of AI blow, where are you setting your sails to go?

Chapter 3
Robotics

The field of robotics stirs up a mix of emotions. To some, it is an exhilarating arena to watch. To others, it is a prospect viewed with dread.

There have been a number of blockbuster sci-fi movies that have portrayed robots as fun, friendly, intelligent, and interactive machines. Other films, however, have depicted them as ferocious and merciless soldiers. Both these types of movies have somehow shaped the public's perception of robots.

In order to balance that impression, let's do a reality check. Let's see where we are with robots today, then discuss where things can possibly go from here.

Industrial Robots

Robots have been in existence for many decades. The US manufacturing industry had its first robot in 1959 at a General Motors facility in Trenton, New Jersey. The robot was named Unimate #001 by US robotics pioneers Joseph Engelberger and George D. Devol.

The story goes that the two gentlemen met at a party in 1956 and discussed Isaac Asimov's philosophies on robotics.[60] From that point on, they decided to work together resulting in the creation of the prototype Unimate robot and many other subsequent projects.

From 1960 onwards, industrial robots have been used to perform mundane as well as laborious and dangerous tasks.

Tasks that may be hazardous to humans—like painting, spot welding, arc welding, palletizing, materials handling, hazardous materials removal and disposal, nuclear decommissioning, as well as assembly—are now performed by industrial robots.

Kawasaki Robotics[61] is one of the world's leading suppliers of industrial robots and robotic automation systems. According to their website, they have shipped more than 200,000 robots worldwide. They have delivered robots and robotic automation solutions to numerous factories, industrial plants (fertilizer plants, cement plants, acrylic fiber plants, etc.), automotive manufacturing plants, metal manufacturing plants, plastic and rubber plants, electronics plants, food and beverage plants, and foundries.[62]

While Kawasaki Robotics is a leading supplier in this field, it has numerous competitors like Mitsubishi Electric[63] (Japan), FANUC Robotics[64] (Japan), Epson Robots[65] (US), Comau Robotics[66] (Italy), Durr[67] (Germany), and Universal Robots[68] (Denmark), among many others.[69]

Industrial plants are not the only ones benefitting from the use of robots. Even poultry farms today are seeing the value of these intelligent machines.

On November 10, 2017, the *Alltech* (alltech.com) website published an article titled "Flocking to digital: The future of poultry technology." Below are excerpts from that article:

> Poultry houses require nearly constant attention—cleaning and sanitizing, collecting eggs and checking birds. This is time-consuming, monotonous work, but it would not bother a robot. Additionally, robots are more precise, thorough and honest about the work they do compared to their human counterparts.

> ...
> France-based Octopus Robots designs entirely autonomous robots to prevent and control disease and infection in poultry houses. The bots also evaluate environmental factors such as temperature, humidity, carbon dioxide, ammonia, sound and brightness.
> Another French robotic company, Tibot, explains that robots can discourage chickens from laying eggs on the floor and also keep the birds moving for an added health benefit. These attributes can result in cost-savings for producers in product and labor while appealing to welfare advocates.
> For more specialized tasks, including feeding and monitoring, Metabolic Robots designed robotic feeders that can increase feed efficiencies, lower mortality rates and alert the producer to potential disease concerns. "Nanny robots" are used by Thailand's Charoen Pokphand Group (CP Group) to maintain healthy flocks of around 3 million laying hens. If the robots detect an ill bird, humans are alerted and the bird is removed immediately. These automations will reduce outbreaks of bird flu and foodborne illnesses, improving the safety of the entire supply chain from producer to consumer.[70]

What is even more fascinating is that companies, such as the Norwegian company Sintef,[71] are developing robots that are capable of extracting breast fillets from chickens. Imagine, what used to be manual processes in food production will eventually be (if they are not already) performed by robots.

What else can robots do?

Service and Hospitality Robots

By definition, the service industry covers any type of business that provides a service or a set of services rather than a product. It is a superset of industries, one of which is the hospitality industry, among many others.

The hospitality industry further encompasses a wide number of fields including lodging, transportation (air, land, sea), and even theme parks.

By the very nature of businesses in the service industry, this domain employs hundreds of thousands of people—serving as airport porters, hotel front desk clerks, hotel luggage clerks, hotel concierges, waiters, cashiers, cooks, cleaners, washers, hotel assistants, hospital assistants, linen clerks, and many others.

In recent years, hotels and hospitals have begun utilizing robots to perform menial and repetitive tasks thus far done by humans.

On February 24, 2018, the *Los Angeles Times*[72] published an article titled "New San Gabriel Sheraton offers robots and high-end food to attract Chinese tourists." Following is an excerpt from that news bit:

> The Sheraton Los Angeles San Gabriel, a 288-room hotel on East Valley Boulevard, has 19,000 square feet of indoor and outdoor meeting space, plus eight robots that can offer guests directions and deliver towels and other items to the rooms.
> In addition to a steak house, the hotel will be home to the nation's first Ba Shu Feng, a high-end Sichuan restaurant that is popular in China. To communicate with guests, the hotel has an account with WeChat, the social app with more than 980 million monthly users.

Japan is in no way lagging behind this trend. On January 20, 2019, *Fox News*[73] published an article titled "A Japanese hotel is going to swap robots for humans."[74] Here is an intriguing excerpt:

> A hotel chain in Tokyo is trying to be the first run by robots, albeit with a few malfunctions along the way, according to a new report.

> Japanese travel company H.I.S. has been opening hotels where robots man the front desk, check-in is handled by a kiosk, and face recognition opens the door to your room.
> The hotel name, Henn na Hotel ("henn na" means strange in Japanese) underscores the eerie presence of robots and the relative dearth of humans.[75]

China is also not to be left behind. On October 21, 2019, *Business Insider*[76] released an article titled "Chinese e-commerce giant Alibaba has a hotel run almost entirely by robots that can serve food and fetch toiletries…."[77]

Recognizing the potential of service robots, other giant hotel chains have also joined the bandwagon.

Hilton has their concierge robot named *Connie.* Aloft uses butler robots called *A.L.O.* Crowne Plaza has delivery robots named *Dash.* While Yotel Hotel has luggage-carrying robots called *Yobot.*[78]

As for robots in hospitals, some forward-looking medical centers began leading the way in using such machines years ago. They have now earned a title: medical robots.

In 2015, the University of California San Francisco Medical Center at Mission Bay deployed 25 robots to transport meals, supplies, and other materials around the hospital.[79]

On June 29, 2016, an article appeared on *The Medical Futurist* website listing the following exciting facts about medical robots. Here are some of them:

- Medical robot *Xenex*, manufactured by Bryan Health, is used to disinfect hospitals. The Westchester Medical Center reported a 70% drop in ICU Clostridium difficile or C. diff, which is a bacterium that can cause a range of symptoms from diarrhea to colon inflammation.
- Two hospitals in Belgium will use humanoid medical robots called *Pepper* as hospital receptionists.

- A medical robot named *TUG* can carry medication of up to 400 kilograms (880 pounds).
- A Japanese medical robot called *Robear*, as it resembles a big gentle bear, can lift and carry patients from bed to wheelchair, and vice-versa.[80]

How many of these medical robots are now deployed?

For one, Aethon,[81] the manufacturer of the TUG delivery robots, has 425 TUG robots deployed in 130 hospitals, mostly in the US, and a few in Denmark, Germany, and Australia. Aethon is not alone in this market. Among its competitors are Panasonic, Swisslog, and Vecna.[82]

However, the use of robots in hospitals is not limited to the delivery of food and supplies.

Stevie, a robot created by an Irish company, Akara Robotics, can tell stories, organize sing-alongs, and call out bingo numbers in a group environment. During the coronavirus pandemic, Akara Robotics also drew plans for a new robot called *Violet* that would be able to disinfect areas using ultraviolet light.[83]

Announcing a giant leap in the research and development of medical robots, *ZDNet*[84] released an article on October 11, 2019 titled "Nurse robot set to make the rounds at major hospitals." An excerpt below tells the story:

> A nurse robot named Moxi is officially coming out of beta testing. Last year, Moxi concluded its first real-world trials to test a collaborative automation integration in a working medical facility.
>
> Moxi was designed by Austin-based Diligent Robotics, an automation firm bringing its hospital robot assistant to the real world starting with a Texas hospital. The company has received backing from investment firm True Ventures and had early financial support from the National Science Foundation, which

awarded Diligent with a series of Small Business Innovation Research grants.[85]

The development of nurse robots might be concerning to a lot of people, especially those who might be displaced by this technology. However, the 2019 coronavirus (Covid-19) pandemic taught us that healthcare providers such as nurses and doctors are at great risk when providing care to patients with highly contagious diseases. While nurse robots may be deemed as less sociable than human nurses, the former will not die from viruses or other deadly microorganisms.

From the above examples, it is evident that service robots, hospitality robots, delivery robots, medical robots, or whatever name you may want to call them, will find their way into mainstream hotel and hospital business operations. The rate of adoption is, in fact, likely to accelerate due to the recent pandemic.

Over the next few years, if these robots prove to be economical, reliable, and safe, then the next question is this: How many hotel and hospital jobs will be displaced by robots?

Social and Eldercare Robots

In June 2019, I attended a robotics conference in Brooklyn, New York. One of the speakers was asked, "What is the next biggest opportunity in robotics?"

"Eldercare robots."

Robots that can help clean the house, wash dishes, converse, give reminders, play games, sing songs, recite a poem, read a book, and even call 911 in an emergency.

Why are eldercare robots going to be the next biggest thing in robotics?

The answer is simple: the median age of the US population has been climbing since the 1960s.

According to the Statista website, the median age in America in 1960 was 29.5 years. In 2018, that median age climbed to 38.2 years.[86]

Why is that?

On June 20, 2019, Bloomberg released an article titled "Half of Americans Are Now Over the Age of 38." The article cited that the median age in America had reached 38.2 years, based on 2018 statistics. This excerpt outlines the trend:

> The national median age rose to 38.2 years last year, according to newly released data from the U.S. Census Bureau. That suggests half of the U.S. population is more than 38-years-old and half is younger. By gender, the median for woman is 39.5 while for men its 36.9 years.
>
> The aging of America is widespread with 49 of 50 states showing an uptick in older residents. North Dakota was the only state to see a decline in its median age, from 37 years in 2010 to 35.2 in 2018.
>
> "This aging (phenomenon) is driven in large part by baby boomers crossing over the 65-year-old mark," said Luke Rogers, the Chief of the Population Estimates Branch at the Census Bureau.
>
> A decline in birth rates is another reason. The number of U.S. births fell to the lowest level in 32 years in 2018.[87]

In other words, the baby boomers (born between 1946 and 1964) in America are now crossing 65 years of age while the country's birth rate is simultaneously declining.

For purposes of our discussion on eldercare robots, the more pertinent information is that the baby boomers are aging.

Those born in 1946 will be 79 years old in the year 2025.

The sad reality—not just in the United States but in other countries as well—is that the majority of the elderly live alone. As a result, they are more prone to accidents, depression, and suicide.

Many old people turn to pets for companionship. However, pets do not come without added responsibilities. Pets can't simply be left in the house or apartment for weeks. They need fresh air, exercise, and to be able to do their daily routine outside the house—whatever the season may be.

While pets are adorable, they cannot help clean the house, communicate like humans, or call 911 in the event of an emergency. For these reasons, eldercare robots offer an option that will benefit millions of seniors in the not-so-distant future.

Japan is certainly one country that has realized the value of eldercare robots.

On March 27, 2018, Reuters released an article titled "Aging Japan: Robots may have role in future of elder care." Here is a section of that article:

> Robots have the run of Tokyo's Shin-tomi nursing home, which uses 20 different models to care for its residents. The Japanese government hopes it will be a model for harnessing the country's robotics expertise to help cope with a swelling elderly population and dwindling workforce.
> Allowing robots to help care for the elderly—a job typically seen as requiring a human touch—may be a jarring idea in the West. But many Japanese see them positively, largely because they are depicted in popular media as friendly and helpful.[88]

Obviously, Japan will not be the only country that will need eldercare robots in the coming years.

According to an article on *Synced* (syncedreview.com) published on June 1, 2019, "The AI in the elder care market is expected to exceed US$5.5 billion by 2022, and will grow into one of AI's most important support roles in societies of the future."[89] The same article mentioned that almost one-fourth of the world population will be aged 60 years and older by the year 2050.

While there is a lot of research and development underway in the area of eldercare robots, some robotics companies are developing multi-purpose robots called social robots.

As I mentioned in my first book *A.I. Hacked*, perhaps the most famous social robot or humanoid today is *Sophia*, a creation of Hanson Robotics based in Hong Kong. The founder of Hanson Robotics is David Hanson, a former Disney Imagineer.[90]

As of this writing, Sophia has publicly demonstrated the ability to smile, perform lifelike head and arm movements, converse, and even sing. It is also widely known that Sophia is the first robot to be granted citizenship by Saudi Arabia.

Eldercare and social robots are foreseen as part of the "new normal." Paradoxically, the gap in human relationships brought about by technology will be filled by another technology: intelligent machines.

Robots and AI in the Military

Robotics and AI will play significant roles in defining who will be the superpower by the end of this decade.

As it stands today, the United States and China are the leaders in AI research and development. While the US is clearly leading the race, China is determined to surpass America by 2030.

This duopoly can be broken by Europe, but it will require a concerted effort amongst the EU member nations—not to mention the huge cost it will entail—to become a major player on the AI global stage.

Back in the United States, one of the most prominent creators of military robots is a company called Boston Dynamics. This robotics company based in Massachusetts—in fact, a spin-off from the Massachusetts Institute of Technology—is now a wholly owned subsidiary of a Japanese company called SoftBank Group.

Boston Dynamics is the maker of *BigDog*, a quadruped robot (having four feet) that was designed for the US military. The research and development of BigDog was funded by the Defense Advanced Research Projects Agency, also known as DARPA.

Boston Dynamics has also created a commercial "lightweight version" of BigDog called *Spot*.

On November 25, 2019, Boston's NPR news station WBUR (wbur.org) released an article titled "Mass. State Police Tested Out Boston Dynamics' Spot The Robot Dog. Civil Liberties Advocates Want To Know More." Below is an excerpt from the article:

> Massachusetts State Police is the first law enforcement agency in the country to use Boston Dynamics' dog-like robot, called Spot. While the use of robotic technology is not new for state police, the temporary acquisition of Spot—a customizable robot some have called "terrifying"—is raising questions from civil rights advocates about how much oversight there should be over police robotics programs.[91]

It is equally interesting to note that a Norwegian oil rig has found business value in Spot.

On February 11, 2020, Bloomberg released an article titled "Boston Dynamics Robot Dog Goes on Patrol at Norwegian Oil Rig."[92]

It is likely that there will be an uptick in the use of robots to monitor and patrol buildings, perimeter walls and fences, as well as other critical facilities in the future.

Boston Dynamics is also widely known for developing a robot called *Atlas*, a humanoid capable of "human-level agility" such as running, walking, jumping, skipping, lifting objects, and even performing some acrobatic stunts.[93]

It is probable that Atlas will one day be trained to be a soldier for warfare. However, you can imagine the amount of testing that needs to take place to ensure that nothing can go wrong with the algorithms controlling Atlas.

Speaking of the military, the United States and China are not the only countries that have determined the criticality of artificial intelligence in building their next-generation defense capabilities. The United Kingdom is on its way.

On January 20, 2020, *Forbes* published an article titled "How Artificial Intelligence Will Make Decisions In Tomorrow's Wars." Here is an extract from that news bit:

> The MoD's Defence and Security Accelerator (DASA) has announced the initial injection of £4 million in funding for new projects and startups exploring how to use AI in the context of the British Navy. In particular, the DASA is looking to support AI- and machine learning-based technology that will "revolutionise the way warships make decisions and process thousands of strands of intelligence and data."
>
> In this first wave of funding, the MoD will share £1 million between nine projects as part of DASA's Intelligent Ship—The Next Generation competition. However, while the first developmental forays will be made in the context of the navy, the U.K. government intends any breakthroughs to form the basis of technology that will be used across the entire spectrum of British defensive and offensive capabilities.[94]

The next-generation defense strategies will be made up of many different military capabilities covering air, sea, and ground.

For air defense, it is widely known that military drones are now being used in tactical operations. Autonomous jets are likewise being developed, if not already undergoing testing.

For naval defense, self-driving ships are a logical next option. Intelligent war ships with fleets of AI-enabled jets and drones will likely form part of a robust naval defense strategy. Furthermore, intelligent robotic fishes and crustaceans can prove to be useful for naval surveillance activities.

For ground defense, robotic soldiers, autonomous tanks, and self-driving military vehicles will likely be critical elements in the army of the future. In addition to the military's ground capabilities, robotic "pets" such as robot dogs, robot snakes, and robot spiders can easily be deployed to places too difficult or dangerous for humans.

It is not implausible to imagine that such advanced technologies can also be used for espionage, surveillance, and countersurveillance. That said, robotics and AI may, in fact, usher in a new Cold War era.

It is highly probable that future conflicts will involve fewer humans. Instead, they will be wars of robot soldiers, autonomous jets, surveillance drones, self-driving war ships, intelligent tanks, and robotic patrol pets.

It sounds like a plot of some futuristic sci-fi movie; but it is not.

It is now more probable than ever.

It will be the new reality.

And yes, reality is stranger than fiction.

The Big Picture

As we have discussed in this chapter, robots will deliver value in many areas of life in the coming decades.

Due to the aging world population, eldercare robots, social robots, and household robots will likely become practical elements in our daily existence. These robots will not only provide companionship to people but may also save the lives of many.

Medical robots, such as delivery robots, disinfecting robots, and nurse robots, will help minimize risks to human healthcare providers and the general public in the event of a pandemic.

Furthermore, robotics and AI will create new job opportunities that will require new skills and training. One can expect new opportunities in the areas of strategic planning, design, production, marketing, and post-sales maintenance and support of products and services related to AI and robotics.

On the other hand, robots used in manufacturing, service, and hospitality industries will displace laborers. Big consulting companies have been predicting that these industries will be hit hard by robotics and automation. It is not a question of if, but when.

The practical approach is to determine what areas will likely be disrupted and then plan a transition.

For the young generation, the key is to educate them on career opportunities in AI and robotics. It will be equally important for young people to be made aware of the potential impact of AI and robotics on various industries, so that they can make informed decisions when setting their career goals.

For those who are already in the workforce that may be affected by these technologies, it is imperative for them to start upskilling for a career shift.

Most job transitions will involve training; but that will be easier said than done. People will need to make time to attend training programs to acquire new skills. They will also have to save money in order to afford such training.

Will the government be able to provide financial assistance?

The more fundamental question is what skills and training should the workforce acquire?

Who will be in a position to point them to the right direction?

On a much bigger scale, will there be enough jobs?

How will workers be able to effectively compete for those jobs?

The harsh reality is that every individual will have to look after himself and his family. The responsibility falls on each one of us. Every person needs to know where the world is going. We can choose to either surf the big wave or avoid it.

The ultimate question is this: *How are you preparing for the future?*

Chapter 4

AI in Finance

One fine day in the autumn of 2019, a friend of mine in Manhattan came to me with palpable excitement. He explained that he had just come back from a high school reunion in China and was eager to show me a technological breakthrough that they had implemented in China. It is called WeChat.

WeChat is a mobile application developed in China and used by the majority of mainland China's residents for social media, messaging, and mobile payment.

My friend explained that he can send money to any WeChat user around the world. He can even order and pay for items through this same mobile app. He called WeChat his "electronic wallet."

With over a billion Chinese people who are potential users of WeChat, this mobile app is a remarkable step towards a cashless world.

Clearly, many countries are likely to be wary about using WeChat. So they will probably develop their own mobile app solution, if they haven't already.

WeChat, then, is just one of the many examples of electronic financial transaction processing emerging globally.

Broadly speaking, the world is moving into online marketing, online ordering and sales, online banking, online payments, and many other forms of online transactions.

As more and more financial transactions become digitized, it is important to note that the digital ecosystem is where AI lives and breathes.

That said, let us now look at how AI is being applied in the finance industry.

Stock and Forex Trading

Stock market. Some of you who have been buying and selling stocks for many years know that trading used to involve calling a trader over the phone and placing your order to either sell or buy stocks for you.

For those who may not be familiar with stocks, let me take a few moments to explain what they are.

What are stocks?

To answer that question, we first need to understand a few basic things.

One, a company needs money to operate and run its business.

Two, money, funds, or capital can either come from the company's private owners or incorporators or it can come from the public.

How can funds come from the public? A company will need to apply for listing in a stock exchange, through which it can sell shares of the company to the general public.

In simple terms, stocks are shares of the company that are sold to the public. That means that, if you bought some shares of a publicly listed company, you own a part of that company. If you own a significant number of shares, you can even have a seat in that company's Board of Directors.

How are stocks or shares sold and bought? Stocks can be bought through stock trading companies or, more recently, through electronic stock trading platforms. Whatever channel you decide to use, each one will charge a fee for every transaction, whether you buy stocks or you sell stocks. While the fee charged by a stock trading company or an electronic trading platform is a tiny percentage, it can easily become a sizable amount if you are buying or selling a large number of "high value" shares.

The value or price of a share is largely driven by market forces; meaning, supply and demand. Because there is a finite number of shares of a company that is available for trading in the stock market, the price of a share can go up or down, depending on how many people are willing to sell or pay for that share.

For a seasoned stock trader, that "willingness to pay for such stock at such price" is mainly based on two key factors: 1) the company's current financial position—profit, market reputation, among other things, and 2) the company's projected growth in the next five to 10 years.

It is the responsibility of the executive team of that company to provide relevant financial information to the general public.

Now let's go back to our main topic, AI.

How can AI then be applied in stock trading?

As stock trading today is purely electronic, AI can be used to constantly monitor stock prices and automatically buy or sell shares for you depending on a set of conditions or parameters you specify.

There are hundreds, if not thousands, of stocks whose prices are moving up or down almost every second. It is not possible for any human being to capture all the voluminous and rapidly changing data points to promptly decide on when to buy or sell stocks. However, AI can ingest all that information, decide on what to do based on some decision criteria, and even execute the sale or purchase at speeds surpassing human agility.

Here are four examples of how AI is shaping the stock trading industry.

Epoque is a company that provides an AI-powered trading system. The company is based in Switzerland. Epoque offers a completely automated AI trading system that is made up of three major sub-systems: 1) an analytical sub-system or engine that monitors and analyzes potential trades, 2) an automated ordering sub-system that places orders and performs numerous operational tasks, and 3) a logical sub-system or engine that handles orders in real-time and utilizes machine learning to continuously improve performance.[95]

Auquan is a London-based company that was recently named the "Hottest Fintech in Europe" at the 2019 Europa awards. Auquan's platform allows data scientists to develop trading strategies to solve numerous investment challenges. In other words, their clients can benefit from data analytics without hiring their own in-house experts.[96]

Back in the United States, a San Diego-based company called Trade Ideas (trade-ideas.com) is making headway. Trade Ideas has an AI-powered trading platform named *Holly*. This platform that uses machine learning "subjects dozens of investment algorithms to more than a million different trading scenarios to increase the alpha probability in future sessions. Only those strategies with a success rate of 60% and above and a 2:1 profit factor are shared with traders the next day."[97]

Finally, there is Arbitrage Technologies, a company based in San Francisco, California. Below is an extract from the website of Welcome.AI, the largest AI professional network, on Arbitrage Technologies:

> Arbitrage Technologies builds next generation trading systems to connect and interconnect financial exchanges on a global level.

> We simplify trading life cycles to give traders and investors - both institutional and individual - access to the best available price in the market and the ultimate position in financial exchanges at any time.
> When Technologies can Arbitrage it means you are among the fastest participants in the market. We create for our customers the possibility to add new revenue streams and to lower their trading fees while not being frontrunned.[98]

Aside from stock trading, AI has made its way into foreign exchange trading, futures trading, and even ETFs (Exchange-traded funds). There are AI-powered software solutions for these offered by companies like VantagePoint Software.

Below is an excerpt from the company's website:

> Created by world-renowned trading software pioneer Louis B. Mendelsohn, VantagePoint forecasts Stocks, Futures, Forex, and ETFs with a remarkable proven accuracy of up to 87.4%. Using Artificial Intelligence, VantagePoint's patented Neural Network processes predicts changes in market trend direction 1-3 days in advance. With this information, traders can confidently get in and out of trades at the optimal time. With nearly four decades and more than $10 million dollars of research and development invested, VantagePoint helps traders preserve their hard-earned capital and create real wealth.[99]

As you can see from the above examples, AI is transforming the stock, futures, forex, and ETF trading industries. Highly paid analysts and traders are now being supplemented, if not replaced, by AI-enabled trading platforms.

AI machines are tireless. They are capable of analyzing large volumes of data. These AI trading platforms can monitor thousands of data points as they dynamically change throughout the trading period. They can place orders far faster than humans possibly can.

The signs are written on the wall. Give it a few more years, stock market analysts and traders will lose their jobs to automation.

In regard to investors, the prudent ones who will use these AI trading platforms will have a distinct advantage over those who do manual trading.

Man and machine is better than man alone.

Insurance

There are natural risks that are inherent to our daily life. In some cases, we have control; in others, we don't. Then there are what we call *acts of God.*

There are various ways to manage a risk.

First, you can ignore or avoid the risk; meaning, brush it aside and pay no attention to it.

Second is to accept the risk. This means that you acknowledge that there is a risk, but you will do nothing to address it. It is like saying, "I know there is a risk, but I will just accept it. If it happens, then it happens."

The third option is to mitigate the risk. This means that you will address the risk to either minimize or eliminate it.

The fourth option is to transfer the risk. How do you do that? You buy insurance to cover the damages in case the risk event happens.

A good example of this is car insurance. You acknowledge that there is a possibility that someone will hit your car. You don't know how, when, or where. But in case that happens, you don't want your hard-earned savings to cover for the entire cost of the repair or replacement of the car.

The basic principle of insurance is simple: premiums collected from a group of insured individuals or companies will be used to pay for claims made by a few. Every insurance company has a team of actuaries and data scientists who calculate the probability of a risk in order to determine the premium payment appropriate for the insured risk.

Calculating risks involves a number of activities and a significant amount of data points. It requires data gathering, analysis of historical information, and calculating risk probabilities. If the quality of data is low, then the risk calculations will not be reliable.

How, then, can AI be applied in this situation?

There are at least three major areas where AI will be beneficial to the insurance industry.

First, automated claims processing and fraud detection. Insurance claims are now being digitized, meaning, these are either submitted through electronic forms or paper-based claims are transformed to digital format through optical character recognition. That said, insurance companies can build AI-enabled claims processing systems. In short, AI can be trained to learn claims processing rules and fraud detection strategies and apply them to every claim that AI reviews.

Second, policy pricing based on customer behavior. This means that the pricing of an insurance policy will be dynamic (i.e., will change every pre-defined period) based on data gathered on the person's driving behavior. In fact, there are driver behavior monitoring systems available today that can provide real-time information to management of trucking or taxicab companies—or insurance companies—on the driving behavior of the person at the steering wheel. Due to a large number of data points that insurance companies handle, AI can be trained how to monitor driver behavior and dynamically assess the risks leading to adjustment of premium charges or even cancellation of policy, in case certain pre-set thresholds are violated by the driver. This information gathering on driver behavior will become even more prevalent when Internet of Things (IoT) is implemented nationwide.

Third, personalized and interactive insurance. The traditional model in insurance is that you call an insurance agent or go online to apply for insurance. Then, the insurance company will review your information and decide whether to approve or decline your application. With the use of AI, on-demand insurance can be made possible. This means buying insurance can be as simple as responding to a chatbot using your mobile phone or laptop. AI will then be able to determine your geographic information, review your insurance history and other personal data, dynamically calculate the risk, and then provide you a quote.

Integration with other systems such as social media platforms will likewise enable an AI-enabled insurance system to proactively suggest coverage for an upcoming event, say, a vacation in Europe.

Banking

The banking industry has become an integral part of the lives of millions of people around the world. Direct deposits from employers. Mortgage applications. Credit and debit cards. Deposits and withdrawals. Automated teller machines. Online and mobile banking. Financial advisory.

The banking industry, in general, has cautiously moved to digital ways of transacting business. Today, it is safe to assume that people prefer to deal with banks that have digital and mobile platforms—meaning, customers can access their bank accounts, transfer funds, or pay their credit card bills using a laptop or a mobile app.

As more and more banking processes become digitized, the more probable it will be for AI to be applied to those electronic processes.

In what areas will AI be beneficial?

There are at least three areas where AI can transform banking.

First, personalized customer service. Customers today demand not only efficient and high-quality service, but also personalized service that is accessible 24x7. Service availability is especially important for bank customers in this day and age. What's more, people are now demanding service that is tailored to their rapidly changing needs.

Banks are now investing in AI to deliver what is called front office conversational banking. This includes seamlessly identifying and authenticating customers, employing chatbots and voice assistants that mimic bank personnel, and providing personalized recommendations.

AI can also be applied to deliver personalized loans processing. To be more specific, while the standard procedure today is to go to a bank to apply for a loan, the bank that will have an AI-enabled interactive loan processing system will attain competitive advantage. This system will not only cover mortgages, but also small loans that customers can quickly obtain to pay an unforeseen expense. This interactive loan processing system can be as simple as a chatbot that will ask a few questions, dynamically assess the risk profile of the loan applicant, and decide on the application within seconds.

A second area for AI in banking is error and fraud detection. Banks largely operate in standardized, consistent procedures designed to detect, if not eliminate, errors. These procedures are also formulated to strictly comply with laws and regulations such as anti-money laundering (AML) and know-your-customer (KYC) regulatory checks, among others. This systematic framework also covers the security of the transaction processing itself and, of course, the transaction data.

However, due to the inherent nature of human beings being prone to errors, there are risks to the traditional system. In order to adhere to the highly regulated environment of the banking industry, banking systems can be enabled with AI to quickly detect deviation from standard procedures and detect possible fraudulent transactions. The AI system can then immediately send an alert to a banking manager to correct these.

A third area for AI, financial advisory services. This is one aspect that a lot of bank customers will find value in. Those who may not be adept in cashflow management, for example, will benefit from an AI engine that can give them practical recommendations. Such an engine can help optimize their cashflow, suggest other relevant banking services, help avoid penalties, or even recommend a schedule of payments to improve a customer's credit score.

Intelligent Automation

Intelligent Automation, often referred to as IA, is an integrated solution for digital transformation by way of combining business process management (BPM), robotic process automation (RPA), and some added AI such as machine learning and natural language processing capabilities.

What is the target market for intelligent automation?

Many companies today embark on a Digitalization Program to enhance their users' experience. They achieve this goal through streamlining business processes, automating procedures, and embedding AI interventions such as chatbots to personalize services.

This means that intelligent automation can be applied across a broad range of office functions: marketing, customer service, IT, accounting, finance, HR, payroll, procurement, and business operations.

As an example, let's discuss IT.

You can apply intelligent automation at a service desk where chatbots can interact with users who want to report a problem. If the reported issue already has a script or a set of documented procedures to resolve, these steps can be automated using a software with machine learning capabilities. Otherwise, the chatbot can refer the issue to a technical support specialist.

Similarly, for HR, you can have a chatbot that will interact with an employee who has an HR-related question or a request. Again, if there is a pre-defined step-by-step procedure to address that employee's question or request, then that procedure can be automated using a piece of software.

You can imagine that intelligent automation can be applied in many other industry verticals, including finance, higher education, real estate, construction, transportation, logistics, and many more.

This new wave of digital transformation across many industries will usher in a more pervasive level of automation in the workplace. However, it will also challenge traditional thinking on tasks performed by humans.

The Big Picture

Artificial intelligence inhabits the digital world. That digital cashless universe is rapidly expanding as more and more companies around the globe digitize their processes and data.

Equally interesting is the rise of mobile applications now fueling a mass exodus from cash transactions to digital commerce.

Years before China decided that they want to be the world leader in AI by 2030, Allen Zhang[100] in 2011 created a mobile application, WeChat. Zhang's app enables its users to not only connect with each other via social media, but also electronically pay for purchases and even transfer funds seamlessly. With over a billion potential users in China itself, WeChat is a giant leap into a cashless world.

In the corporate arena, there has been an acceleration in the transformation of financial data to digital format over the last 10 years. Furthermore, as a consequence of the emergence of AI technologies, major financial institutions like banks, stock trading companies, and insurance firms have recognized the business value of AI: efficiencies, cost reduction, data accuracy, service availability, and even security and fraud detection.

However, all these benefits to business come with a price tag.

In October 2019, Wells Fargo predicted that over 200,000 jobs will be lost due to automation technology over the next decade.[101]

Even highly paid Wall Street employees are at risk of losing their jobs due to the rise of AI platforms that can outperform humans in data analysis and stock trading.[102]

The more widely financial institutions implement automation technologies, the higher the probability that human workers will be displaced.

At the same time, as more financial transactions become automated and more traceable through the use of blockchain technology, there will remain to be unscrupulous elements in society who will try to break into financial systems. Such criminals will become more sophisticated. Financial crimes will likely involve highly advanced technologies. That said, there will be a growing need for more information security specialists to protect financial systems and data.

As a result, governments will tighten financial controls and will likely implement some form of electronic ID system to allow for speedy access to information and traceability of financial transactions. For one, the Chinese government is known to have already implemented facial recognition in many parts of their country.

In October 2019, France announced that it is planning to use facial recognition as a mandatory digital ID for its citizens.[103] Now that the French government will be taking this step, it won't be long before the rest of Europe follows.

Another consequence of automation is that nations will likely tighten immigration and give preference to highly skilled workers over human laborers whose tasks can be performed by AI-enabled solutions.

In fact, on February 18, 2020, the *BBC News* announced to the world, "Immigration: No visas for low-skilled workers, government says." The news bit explained that, as a result of Brexit, the UK government is urging UK companies to not rely on "cheap labour," but instead develop automation technology.[104]

As we march into the future, the story will unfold: AI will enable financial institutions such as banks and insurance companies to provide efficient, personalized, interactive, and highly available service to their human customers—using fewer human workers.

The trajectory may be disturbing to many, but the probability that these predictions will happen cannot be ignored.

Logic dictates it is just a matter of time.

Chapter 5
The AI Economy

Economy is a broad term that touches almost all areas of life. The condition of a country's economy affects people's health, welfare, shelter, food and nutrition, education, jobs, safety and security, transportation, communication, among others.

The economic indicators used to determine the health of our national economy include the gross domestic product (GDP) growth rate and the number of new jobs generated per month.

The number of new jobs generated per month is a critical metric because it indicates whether businesses are growing, and therefore hiring more people. More importantly, a low unemployment rate means that there are many people who have the means to buy and/or sell commodities and services. That buying and selling will spur growth in other industries and, in turn, will generate more jobs.

However, we have seen from prior chapters that AI and robotics have the potential to displace millions of jobs. To be more specific, laborers who perform repetitive, predictable, procedure-based tasks will likely be eased out by software automation or robots.

Considering that laborers typically make up the majority of workers in a company, the potential impact is immense—both at the microeconomic and macroeconomic level.

Therefore, if one of the indicators of a healthy economy is the number of new jobs generated per month, will our country's economy truly be healthy with increased automation and fewer human workers?

At the ground level, a company operating mostly with robots and AI may deliver huge profits to its owners, but what benefit will that bring to the people in the surrounding community?

In truth, the portrait of a multimillion-dollar company that provides livelihood for an entire city or town is now fading away. Today, there are multibillion-dollar companies that have very few human workers. This concentration of wealth in the hands of a few individuals is not just disturbing. If this trend continues, it will be devastating to the majority of people who are essential in keeping the national economy going.

While one can argue that a robot can be taxed in the same way as a human worker, the latter plays a far more critical role in society than a machine does.

The difference between a robot and a human worker is that a robot will not buy food, shop for clothes, rent an apartment, buy a house, get a haircut, purchase a new laptop, or book a holiday vacation—those human activities that will stir growth in a wide range of industries.

Yes, robots will work 24 hours a day, seven days a week. They will consistently perform the tasks that they are programmed to do. They will not take vacations. They will not get sick. However, the downside is that robots will not spend money as humans do.

That said, while automation will likely displace millions of jobs, it is important to note that the AI and robotics industries will also generate new jobs. There will be additional work opportunities in the areas of strategic planning, design, production, marketing, post-sales maintenance and support, training, among many others.

However, the question is this: Will the new jobs that will be generated by AI and robotics match or exceed the number of jobs that will be displaced by these technologies?

Initial estimates indicate that they will.

According to a research study conducted by the global consulting company PwC, "while AI could displace roughly 7 million jobs in the U.K., it could also create 7.2 million jobs, resulting in a modest net boost of around 200,000 jobs." That PwC research was cited in a news article that appeared on the *CNBC* website on July 17, 2018.[105]

Furthermore, on September 18, 2018, an article titled "Artificial Intelligence To Create 58 Million New Jobs by 2022, Says Report" appeared on the *Forbes* website. The article cited a report from the World Economic Forum (WEF) predicting that automation in the workplace will create 133 million new jobs; but will displace 75 million jobs by 2022.[106]

While the initial WEF forecasts are bright and sunny, one key question remains: How do you equip each of the 75 million people to transition to any of the 133 million new jobs that the AI industry will spawn?

In order for us to answer that question, we need to understand the universe of workers that will potentially be impacted by this change.

The sector of society that will be most affected by AI is the labor sector—the cab drivers, bus drivers, truck drivers, warehouse workers, porters, front desk clerks, among many others. They are the workers who earn low wages and may not have the spare money to spend for training, that is assuming that they even have the time to study after a laborious day. Some of them may even have a second job to make ends meet.

Therefore, it does not automatically follow that the 75 million people whose jobs will be displaced by AI will have a safe landing within the 133 million new jobs that AI will create. Non sequitur.

What is clear is that there is a need for government programs to educate and upskill the people whose jobs will likely be automated. It is not an option, but an imperative; otherwise, the government will end up with millions of jobless people to support.

A number of futurists, authors, and prominent business people have predicted that governments in the distant future will need to provide unconditional income or a universal basic income to their citizens in order for their economies to stay healthy. We have seen a preview of that in the form of a stimulus package during the Covid-19 lockdown in the United States.

This unconditional and guaranteed income from the government might appeal to many people; but the idea is contrary to the fundamental principles of capitalism. People will quickly dismiss this idea of government dole out money as a form of socialism. At least that is the prevalent thinking in this day and age.

However, as AI and robotics mature, the technological innovations that capitalism fertilized and germinated may produce some of the most difficult challenges that capitalism itself will need to address in the future.

Capitalism will be self-defeating if the commodities that the country produces and the services it offers are beyond the reach of its citizens, whose jobs will be displaced. It will be self-defeating because the capitalist economy, or any economy for that matter, needs commerce—buying and selling of products and services—in order to stay healthy. A high unemployment rate can only hurt the economy.

Again, robots can manufacture a wide variety of products and deliver an impressive array of services, but they will not consume those products or services. Therefore, one of the ways to keep the economy going and growing in the future, as predicted by some analysts, is to provide some living means to people; that is, to give each citizen a guaranteed basic income.

As previously pointed out, the other option is for governments to develop effective retraining programs for their unemployed citizens. It is clear that AI and robots will create new job opportunities. Furthermore, AI and robotics will certainly not be the only industries that will be around in the coming decades.

One can only guess that there will be new jobs that we haven't thought of yet today that will surface in the future. That is no different from people during the agricultural era who did not envision that we would one day need agile software developers.

As a practical step into the future, every government will need to develop a national database that will capture information from employers and match this with data on unemployed people. AI algorithms can readily be used to optimize that matching.

That sounds all good and promising.

However, this tug-of-war between humans and machines may, in the long run, turn out to be less favorable to homo sapiens.

So then, what if the world—a century from now—reaches an apocalyptic point that there are more jobs for robots than for humans?

Let us analyze the socio-economic and political implications of that scenario for a moment.

Competition for jobs will be stiff.

Governments will have to subsidize their constituents.

Border security will be tightened.

Immigration will be highly regulated, as governments will subsidize their own people. There will be a preference for highly skilled individuals.

To further control government subsidies, every country will set up its own electronic national ID system using technologies like facial recognition, iris recognition, fingerprint recognition, or body chips.

As automation will be performing most of the work, there will be fewer working hours for humans—which means that people will have more idle time. Perhaps that is a good thing. Some will even surmise that more free time for everyone should be the ultimate goal of any country. Maybe so. However, too much of anything is not good.

It is often said that an idle mind is the devil's workshop. That means that too much idle time in the hands of the general population may unintentionally result in an increase in crime rate. If that happens, then governments will all the more tighten their security controls over the general population by way of implementing more sophisticated surveillance technologies.

Due to pressure from various aspects of life, individuals will experience elevated stress levels; consequentially, there will be an increase in mental health issues. Some believe that as a result of such socio-economic pressure on individuals, as well as economic and political demands on governments, world population will eventually decline.

Then again, nothing is for certain. No one is truly sure what the future holds.

Mankind is entering uncharted territory.

As we watch history unfold, then we will know whether AI—a technology that we ourselves created—is the best thing or the worst thing that could have happened to us all.

Chapter 6

Survival Skills

There have been numerous chapters in human history that brought great uncertainty to entire populations. World wars. Plagues. Famine. Stock market crashes.

There is yet another interlude that is waiting to happen. It is a quiet upheaval that will potentially disrupt life over the next several decades. It deserves our close watch.

This coming passage may be likened to the historic transition from the laid-back agricultural age to the manufacturing era brought about by the Industrial Revolution.

Indeed, with the rise of AI, a major shift is once again imminent. This time, from the manufacturing era to the Age of Automation.

AI will change the job landscape. Therefore, there is a need for a paradigm shift; a need to understand the skills required to survive the future.

The skills and the training needed during the Industrial Revolution will not necessarily apply to the coming age.

In order for us to know how to survive the future, we first need to understand a few things.

One, what AI can do.

Two, what AI cannot do, meaning that gap where humans will still be needed.

Three, what skills are required to fill that gap.

Let's begin.

What can AI do? What are its strengths?

AI can perform repetitive, procedure-based, and logic-driven tasks. It can ingest large volumes of data and correlate them. It has the capability to learn and retain that knowledge for a long time. AI does not grow weary. It does not get sick. It does not need medical or retirement benefits.

What are AI's weaknesses?

AI is not good with adapting to constant change. AI is not suited for unpredictable environments. AI does not have emotions. Compared to humans, it has limited capacity to relate to people. It has no intent, no morals, and no values. So far, AI needs large volumes of data in order to "learn" something that may be easy for humans to quickly comprehend.

Given that brief overview, here are the skills and qualities required for humans to survive the future: creativity, empathy, entrepreneurial and managerial skills, technical skills, and interpersonal skills. What about values?

Now that we know what skills we will need in an AI-equipped world, how do we acquire them?

Chapter 7
Creativity and Connecting the Dots

I loved poetry when I was growing up. I remember being fascinated with how a combination of a few words or phrases could evoke so much feeling and express deep thoughts. Words somehow connect to the seat of my emotions.

Just around the time I reached 12 years of age, I tried to teach myself how to play the guitar. My aunt, upon noticing my interest in music, sat down with me for a few minutes to teach me how to read a guitar chord chart. What followed were hours, days, weeks, and months of practice.

During my free time, which was abundant then, I would play a vinyl record, then try to mimic the guitar playing that I just heard. I simply enjoyed it and didn't mind the time. However, that repetitive simulation must have been annoying to many people around me, including my grandfather who once asked, "Is that music?" I laughed.

I was about two years into guitar playing when a friend of mine said to me, "I know you write good poetry and you can play the guitar. Why don't you put them together to write songs?"

Frankly, all I knew was that I was interested in two distinct hobbies: poetry and guitar playing. I honestly did not see the connection until my friend pointed it out to me.

Since then, I have written numerous songs that have not only been recorded by professional singers but have also won songwriting competitions and even garnered music awards.

However, the journey was much more difficult than it seems. I lost in many songwriting competitions before I started winning. Even after winning one competition, I lost in the next one. It took me many years of writing, countless hours of struggle to find the right words, many revisions, numerous moments of joy as well as some occasional frustration. All for the sake of coming up with what I hoped would be masterpieces. Such is the creative process.

So, what is creativity?

There are three key points to that question.

Creativity is defined as "the use of the imagination or original ideas, especially in the production of an artistic work."[107] It is often associated with three words: originality, invention, and innovation.

History is full of examples of creative minds: Leonardo Da Vinci. The Wright brothers. Michelangelo di Lodovico Buonarroti Simoni, also known as simply Michelangelo, Galileo Galilei, Antoni Gaudi, James Watt, Frank Lloyd Wright, Thomas Edison, Marie Curie, Nikola Tesla, among many others.

One common thread among creative thinkers is their ability to courageously operate—to think and act—outside the bounds of the ordinary. That is why they are called *extraordinary*.

For centuries, humans travelled on land and sea. Around the late 1800s, Wilbur and Orville Wright, also known as the Wright brothers—the pioneers of aviation in the US—got this idea of air travel and transformed that concept into an actual flying machine.

Similarly, when the whole world was content with candlelight, charcoal, bonfires, and fireplaces, Thomas Edison had a novel idea of a light bulb—that is, light generated by a flow of electrons.

These are just two of the many examples of what creativity means. It is not simply copying and pasting. It is not just listening to a vinyl record and playing exactly the same chord pattern. Those activities may be building blocks for creative endeavors, but creativity is much more than just copying a way to play the guitar.

I have learned from personal experience that originality means coming up with your own idea. It is connecting the dots in your brain to produce an original piece that is like no other. It can be a totally brand-new idea, or it may be building upon an existing one. Nonetheless, it is coming up with a new concept, a new approach, a new method, a new style, a new product, or a new service.

So, what is creativity?

The first key point is this: Creativity operates outside the boundaries of the norm. Think outside the box. This is a critical point to understand. Creativity, in most cases, cannot be framed in a linear and highly structured environment. Structure is essential in life; but too much structure can impede creativity.

The second key takeaway is that creativity is mainly about immersion. Marinating your brain with ideas. This simply means that creativity needs a lot of time and focus. I am talking of months and even years of doing and refining whatever is being created. It demands a tremendous amount of concentration. It requires a lot of practice.

The third key point: creativity needs an environment where an individual feels safe to fail and to learn from failure. Creativity needs a secure place for experimentation. A hypothesis will not always meet success at first try. The world may not always be ready for a brilliant idea at the time of its conception. Curiosity is usually a powerful impetus that drives humanity to tread beyond its known limits, and that treading usually connotes uncertainty.

With all that said, how does one develop creativity?

The first step is to understand yourself, your interests, and your motivation. You are the main character in the creative process.

There are three questions you need to answer: What do you want to be? What interests you? What motivates you?

What do you want to be? Do you want to be a physics professor, a chef, a business owner, a scientist, a musician, an inventor, a fashion designer, an architect, a Formula-1 automotive engineer, or an astronaut?

What interests you? Do you enjoy cooking, reading fiction books, woodwork, writing poetry, photography, drawing, conducting scientific experiments, playing musical instruments, analyzing the details of a flower or a plant, snorkeling, or observing different kinds of fish in your aquarium?

What motivates you? Do you dream of being famous? Do you want to be rich? Do you hope to be known as a great inventor? Do you like to be perceived as a leader? How about being known as a caring person? Or a scientist who saved the lives of many people?

The second step is to be decisive. Ask yourself this question: How much time am I willing to devote to develop my creativity? Be intentional. Allocate downtime to let your imagination roam. Set aside time for creative activities. The constant impulse to be busy and stay busy does not promote creativity. The persistent stimuli from electronic gadgets can impede the creative thinking process. You need to deliberately set aside time outside of your normal routine to immerse yourself in your creative universe. Don't be afraid to let go of being busy—in fact, be comfortable with being bored. As others have pointed out, boredom is fertile ground for creativity.

The third step is to take concrete steps to pursue your interests. How do you do that? You read and you read more. You research and gather data. You practice. You experiment. You try this and that. You discuss with other people who have the same interests as yours. You brainstorm. Refine ideas. Improve. The more data, visuals, sounds, actions, and experiences you collect, the greater the chance that your brain will be able to connect the dots to produce a novel idea.

Now that we have some idea of what creativity is all about and how to develop it, the question then is this: Is our current educational system designed to promote creativity?

If it isn't, then how will we be able to capitalize on this human advantage we have over AI?

Chapter 8
Empathy

Artificial intelligence algorithms and robots run on electrons that flow through some circuit boards. They have no emotions. It is not inherent for AI and robots to truly appreciate the concepts of kindness, respect, grace, hope, and love.

Machines have no empathy.

Before we jump into deeper discussion, let us first define the word *empathy*.

What is empathy?

The Merriam-Webster Dictionary defines empathy as "the action of understanding, being aware of, being sensitive to, and vicariously experiencing the feelings, thoughts, and experience of another of either the past or present without having the feelings, thoughts, and experience fully communicated in an objectively explicit manner."[108]

In other words, empathy is the ability to put yourself in the shoes of others.

The Value of Empathy

What is the value of empathy?

Empathy is usually associated with the caregiving profession. Nurses. Doctors. Social workers. Rightly so. In fact, it would be ironic to find a healthcare professional who does not care about how patients feel. After all, doctors, nurses, social workers, and other hospital workers for that matter, are in the healthcare profession to promote and ensure the health and well-being of people.

However, experts have come to realize that empathy plays an important role in many other industries, not only in the healthcare profession.

Case in point, there is something called Design Thinking Process or Design Thinking Methodology wherein the first of five stages is Empathize.

This design thinking process is not only applicable to software development, but also in many industry verticals that involve designing for customers. Therefore, this design thinking process can be applicable to architecture, furniture design, fashion design, automotive engineering, ergonomics engineering, and yes, even robotics.

Why is Empathize the first stage in the design process?

It is very simple. An architect, for example, cannot conceptualize a home for another person without considering his or her personality or knowing what he or she is looking for in a house. Just imagine designing a six-bedroom house with only one toilet and bath. Or a civil engineer planning a building in the northeast of the United States with no heating system. Or an automotive engineer omitting air conditioning in the design of cars that are intended to be sold in countries near the equator, like Singapore, the Philippines, Thailand, Brunei, and Indonesia.

Even service desk personnel in large corporations are trained to empathize with customers.

Imagine calling a service support number and hearing a customer service representative say, "How come other customers aren't experiencing this problem you are reporting now?"

Or consider how you would rate the quality of service of a technical support person in your company who tells you, "I really don't understand why you are angry that your laptop has not been working for the past two days."

This skill called *empathy* is not only critical at the rank and file level; it is equally important or even more critical at the leadership and management levels. Careless words from a leader can demoralize a whole organization, a community, or even an entire nation.

Three Steps to Develop Skills in Empathy

How then do you develop skills in empathy? There are three practical steps that you can take.

Step 1: Understand yourself.

In order to develop your empathy skills, you need to start with understanding yourself—your emotions, your thinking process, your reaction to the words and actions of others.

Why is that important?

Understanding your own self will help you understand other people.

What disappoints you will likely disappoint others as well.

What makes you angry will likely irate others also.

There is a good chance that the words that make you feel respected and valued will make others feel the same way too.

In other words, your understanding of your own self will be the barometer by which you can calculate how others will react to what you are about to say or do.

A person with developed empathy skills will ask this question first before speaking: "How would I feel if someone were to say this to me?"

A problem arises when a person does not have a good insight about himself and his emotions. In this case, the barometer, to begin with, is defective. A person who has no proper assessment of himself will likely get into constant arguments with other people, thereby being considered a difficult person to work with.

For example, a person with narcissistic tendencies, meaning one having an unrealistic view of one's self-importance, will have difficulty understanding the travails of others. A narcissist will be firmly convinced that he is better than everyone else. A narcissist will view any success or accomplishment of his team as solely due to his outstanding leadership skills. On the other hand, any failure of his team can never be his fault.

Step 2: Learn to actively listen.

The second step to developing empathy skills is to learn active listening.

This kind of listening goes beyond just hearing the words from the other person and waiting for your cue. Active listening is intentional; that is, deliberately trying to understand what the other person is saying and what that person is feeling.

In other words, active listening is focusing on the other person, not on yourself. To actively listen means that you listen first, not prepare what you want to say while the other person is still speaking.

Active listening also means showing signs that you are truly interested in what the other person is saying or sharing. You can demonstrate interest by calmly asking clarificatory questions to appreciate the context of what is being said.

Step 3: Learn to validate the feelings of others.

The third step to developing empathy skills is to learn how to validate the feelings of other people.

It is possible that you can actively listen to another person but fail to validate that person's feelings first. In a customer service situation, for instance, your mind may be rushing to quickly resolve an issue or a complaint, while the customer simply wants to hear you say, "I totally understand why you feel frustrated. I am sorry that you feel that way. I would feel the same, if I were in your shoes."

Conflicts usually happen when one fails to validate the feelings of the other. Validation of feelings is important as it gives a clear signal to the other person that you care about how he or she feels. To skip that validation step is to subliminally send the message, "I really don't care about how you feel. Let's just discuss how I can solve your problem."

At this point, it is again important to ask this question: Is our educational system today designed to develop our children's empathy skills?

In the final analysis, AI and robots can be trained to respond to humans using some pre-canned scripts. Their audio can be programmed to sound exactly like humans. They may also learn some nuances of the language that they are trained to speak. But at the end of the day, AI and robots are just electronics.

As far as empathy is concerned, humans will still need humans.

Chapter 9
Entrepreneurial and Managerial Skills

While AI's strength is in learning and executing repetitive and procedure-based tasks, it is not yet capable of adapting to an environment where there is constant change; at least not the level of AI that we know today.

If we analyze that thesis carefully, then we will arrive at the conclusion that a significant portion of labor will be displaced by AI, while business owners, executives, and managers may be less affected.

Why?

Because business owners and managers usually deal with constant change.

In my over 30 years of experience as a global IT leader, I can say with certainty that on a daily basis, managers handle a broad range of activities that require making a decision or giving some guidance.

It is not surprising, then, that human entrepreneurs and managers will continue to be much needed in the future. In fact, entrepreneurs, executives, and managers will be the decision makers who will assess what jobs will be automated.

This leads us to the next two essential skills that will be needed to survive the future: entrepreneurial and management skills. They have some similarities, but they have their unique characteristics.

Let's start with entrepreneurial skills.

Developing Entrepreneurial Skills

Entrepreneurship is the ability to transform an idea into concrete action resulting in a product or service. Entrepreneurship requires creativity and innovation. It involves the ability to manage risks. It requires initiative, organizational skills, and the ability to plan and manage projects.

How do you develop entrepreneurial skills?

There are seven practical steps to develop your entrepreneurial skills: 1) Exercise creative thinking, 2) Start a small business, 3) Learn from mistakes, 4) Manage your finances, 5) Practice your leadership skills, 6) Develop and implement a marketing strategy, and 7) Learn from a mentor.

This list is by no means complete, but it is a good start.

Step 1: Exercise Creative Thinking

Entrepreneurs think outside the box. They spot business opportunities that others do not see.

In Chapter 7, we discussed how to develop your creativity skills. I mentioned that you need to understand yourself and the things that interest you. That is your starting point: the things that interest you. Sports? Food? Clothes? Computers? Houses? Books? Cars? Games? Music? Travel?

In this section, however, you take a step further and exercise creative thinking.

What business opportunities are there in the area of your interest? Is there an opportunity to buy and sell? How about creating a new product and selling it? How about offering a new service?

Ask yourself, if you were another person, would you be willing to buy the product you are selling or gladly pay for the service that you are offering?

Explore. Try something new. Visit other places. Expose yourself to different cultures and ideas. Think differently. When the rest of the world is pursuing employment, think of setting up your own business.

Step 2: Start a Small Business

While it is good to dream big, it is prudent to start small when it comes to setting up a business. In fact, a lot of corporate enterprises today started as small businesses.

There are many real-life examples of companies like that one tiny grocery store that mushroomed into hundreds of stores nationwide. Or that small hotel that became an international chain of hotels. Or that small burger joint that later became a global brand.

Whether you plan to do it full-time or part-time, start small, then learn as you grow your business.

Step 3: Learn From Mistakes

Make no mistake, you *will* make mistakes. This is the reason why you want to start small. Human as we are, we are bound to commit errors.

Clearly, the risk is low when you make a mistake in running a small business. Feel safe. Gain experience. Face your errors. Be kind to yourself. Learn from your mistakes in the early years of your business. Strive to continuously improve.

Be systematic in measuring your improvements. How? Review how you are doing on a regular basis. Maybe monthly. Perhaps quarterly. The point is, check if you are improving in managing your business or if you are making the same mistakes over and over again.

Perhaps you need to tweak the design of your product. Maybe you need to simplify your service model. Or perhaps you need to shift your marketing to social media. Whatever the error, forgive yourself, learn from your mistake, and move forward.

Step 4: Manage Your Finances

Business requires capital. It takes money to make money. Regardless of the type of family you are born into, financial prudence is a must when starting or running a business.

You don't want to find yourself buried in debt due to unwise spending. Control your impulses. Be frugal in spending your capital.

A lot of businesses fail not because they are not generating revenue but because they are spending more than what their business is earning.

More importantly, learn how to budget, understand the cash flow of your business, its assets and liabilities, and its profit and loss. You will need that knowledge when your business grows.

Step 5: Practice Your Leadership Skills

You are the founder of your business. You are the owner. You cannot just sit back and let someone run it. There is no such thing as leading by sitting back. You cannot just keep quiet. There is no such thing as leading by silence.

Take the baton and lead your orchestra. Lead planning sessions, even if there are only two of you. Lead the marketing strategy development discussions. Take charge of your projects. Plan your time, resources, and costs.

Leading also means listening to others. Build your business on great ideas. Don't fool yourself into thinking that you have a monopoly over the best ideas in this world. Motivate your team. Learn to be stern when needed. Exercise empathy.

Step 6: Develop and Implement a Marketing Strategy

You have this great idea for a product. You have produced a hundred of them. Or you have developed a brilliant service model. But you don't have a marketing plan. What do you think will happen?

I have news for you. Customers will not miraculously show up at your doorstep. Your product or service will not suddenly become the talk of the town.

You have to market your product or service. You have to develop a marketing plan.

What does your product or service offer? What makes it unique? What is your advantage over competition?

Who is your target market? What age range?

Gender? What geographical location? What income level? Where do they usually spend their time? Malls? Restaurants? Bars? Do they listen to the radio? Do they read newspapers? Do they subscribe to journals? Do they spend most of their time on social media?

You have to plan and budget for marketing. This is one expense that you have to carefully analyze. You need to make sure that you are reaching your target market using the right marketing channels. Otherwise, you will be wasting money on marketing that is not getting you anywhere.

Step 7: Network with Other People

Setting up a business is typically not a solitary or isolated endeavor, unless you are working on a highly confidential project.

As a business owner, you will likely have to deal with suppliers, customers, and other third-party companies. In other words, you will need to interact with other people. The contemporary term for that is *networking*.

Networking will give you varied perspectives, new insights, and fresh ideas for your business. Sometimes you might even get an unsolicited comment that can help you improve your product or service.

Networking will open up opportunities to promote your product, connect with other suppliers, draw inspiration from other entrepreneurs, or learn from the mistakes or success of others.

Interacting with other people will expose you to a wide range of situations that will help you develop your communication skills, interpersonal skills, negotiation skills, persuasion skills, conflict management skills, among others. These are skills that are essential for you to be an effective leader. One day, they will come in handy when you find yourself leading a large organization that you envisioned.

Developing Managerial Skills

An entrepreneur jump-starts a business. However, as the business grows, there will be a need for other people to help manage the company. An entrepreneur will need what is called a *management team.*

Broadly speaking, in a large corporation there are three layers of management. You have the corporate executives or officers, the middle managers, and the supervisors.

The executives are responsible for setting the business goals or what is called the business strategy. Some call it *The Big Picture.* Executives are accountable for managing the finances or the profit and loss (P&L) statement of the company. As such, they can make or break the company. The executives decide where to invest capital; for example, purchase expensive equipment, buy a building, or acquire multimillion-dollar software. They also control the operating expenses of the company, including payroll, building rental, equipment maintenance, supplies, travel, training, among many others.

Middle managers develop the plans and programs that align with the business strategy set by the executives. They work with their team members as well as other departments to successfully deliver each project that they undertake.

Supervisors deal with the rank and file on a daily basis. As they are much closer to the people who perform the work, the supervisors are usually the most technically knowledgeable compared to others in the management hierarchy.

Regardless of level, there are common skills required across the various layers of management: planning, communication, problem-solving, decision-making, motivating skills, and delegation.

Planning

By and large, the success of a business is achieved by way of thorough planning and precise execution. That is not to say that a company will never go through situations beyond their control that require management to make an educated guess. Nonetheless, planning is an integral part of the life of any company—whether a formal or informal planning process is employed.

Generally speaking, there are two types of plans: 1) a strategic plan and 2) a tactical plan.

Strategic Plan. A plan is said to be *strategic* if it enhances a company's competitive advantage, meaning, a company's advantage over its competitors. The word strategic is also usually associated with long-term business goals. Long-term may mean three to five years; in some cases, it means five to 10 years.

For competitive advantage to be a real business advantage, it has to have two characteristics: 1) not easily copied or replicated, 2) sustainable.

What is an example of competitive advantage?

If you own a patent for a product produced by your company, then you have a true competitive advantage as your product will not be easily copied by your competitor. In fact, it will be illegal for your competitor to do so.

Tactical Plan. A plan is called *tactical* if it meets the business objectives for the year, generally speaking. In other words, the word tactical connotes short-term business objectives.

Some examples of short-term objectives may include increasing sales by 5% this year compared to last year or reducing operating expenses by 4% this year.

Regardless of whether it is strategic or tactical, the planning process involves a number of activities: 1) define the business objectives, 2) determine the projects or tasks to meet the objectives, 3) determine the resources needed to implement the projects or tasks, 4) develop the budget, 5) estimate the timeline, then 6) submit the plan for management approval.

In summary, a plan has to have an objective, a set of tasks, required resources, budget, and a timetable.

The truth is, we may have been applying these principles when organizing a party at home, a vacation, or a trip to visit a relative.

Clearly, you don't have to be in a corporate setting to teach your children this valuable skill.

Communication

In my line of work as a senior director in Information Technology, most of my day at work is spent on communicating with my team, my peers, as well as my superiors.

In my thirtysomething years of managing international teams, I have learned that communication involves three important things: 1) listening, 2) organizing your thoughts, and 3) ensuring your message comes across as intended.

Let's discuss listening.

Listening. In Chapter 8 on empathy, we discussed active listening. Allow me to add that in managing people, it is important to listen to the ideas of other people. No manager is perfect. No one has all the great ideas in this world. Listening sends a strong signal to your co-workers that you value them and their thoughts, concerns, and suggestions.

What are the steps involved in listening?

Listening involves hearing what the other person is saying and processing what was said.

Hearing. Listening means intentionally waiting until the other person is done speaking before formulating your response. You are not listening when you are busy thinking about your response. Listening may also involve asking questions to clarify what was said by the other party. An effective technique in listening is to say, "Let me repeat what I thought I heard from you."

Processing. Processing the words that were spoken involves understanding the meaning of the words, the context, and the associated nuances in what was expressed. Is the other person posing a question? Is the question answerable by Yes or No? Is he expressing an opinion? Is the other person soliciting your opinion? Is he citing a fact or hearsay? Is he angry, sad, or disappointed? To whom is the feeling directed?

Now that we have tackled listening, let's discuss organizing your thoughts.

Organizing Your Thoughts. There are multiple ways to organize your thoughts: priority or urgency, chronological, sequential, or logical.

Priority or urgency. Organizing by priority or urgency means that you are listing in your mind the most urgent matter you need to address versus those that are less critical. For example, you need to address the person's feelings first (urgent) before you can explain your proposed solution. Why? The other person will likely not listen to your proposal until you acknowledge his feelings. You can use the same prioritization technique when you are talking to someone who is enumerating a list of items, topics, or requests in your conversation.

Chronological order. Chronological ordering means that you are organizing the information according to the order of time—from the earliest to the latest event. An example of this can be, "At 11AM, Joe called John. At 11:35AM, Joe called Maria. At 12NN, John called Heather."

Sequential order. Sequential ordering means you are determining the general sequence of events. For example, "Joe called John and Maria separately. Then, John called Heather."

Logical order. Logical ordering means that you are applying logic or reasoning in organizing information. For example, "Joe called both John and Maria separately, but the information was obtained by Heather from John, not Joe." Applying reasoning to your thoughts in this way could lead to an insight that you can then convey clearly to others.

Now that you have organized your thoughts, then it is time to ensure that your message is conveyed effectively.

Ensuring Your Message Comes Across as Intended. Remember that your audience cannot read your mind or your intentions. They will only hear or read your words.

There are several tips you can use to effectively communicate your message: choose your words, use proper grammar, seek confirmation.

Choose Your Words. Choosing your words means that you should use language that is appropriate to your audience. Of course, it goes without saying that you have to know your target audience first.

Choosing your words means that you intentionally and prudently select your vocabulary, expressions, forms of address, level of technicality, and so on to fit your readers or listeners.

Talking to a high-ranking official whom you just met for the first time in the same way that you talk to your very close friends may be construed as offensive. Using a wrong pronoun or a wrong title to introduce a person can quickly turn into an awkward situation. Similarly, a presentation to members of a Board of Directors will demand the use of sophisticated business language.

On the other hand, while formal language is appropriate in the corporate setting, it is not necessarily applicable when talking to your teenage children.

Using words that do not fit your audience will make the recipients of your message feel puzzled, unimpressed, offended, or uninterested. Worst of all, they will shut off and not listen to you.

Use Proper Grammar. Getting your message across entails following the rules of grammar, as well as spelling and punctuation. A misspelled word can inadvertently change the essence of your message. The position of a comma in a sentence can make a big difference. Mixing tenses in one sentence will confuse your audience.

While your former high school friends may be forgiving of grammatical slipups, top management in a company will likely be very critical. When applying for a job, a resumé with grammatical errors will not impress a hiring manager.

Even a casual text message or a post on social media that has no commas or periods may not yield a positive reaction.

When speaking to an audience, it is important to pay attention to where you pause in a sentence to ensure that your message is clear. Reading out a paragraph without pausing to indicate a comma or a period could confuse the listeners.

Seek Confirmation. The best way to ensure that your message came across as intended is to seek confirmation. There are numerous ways confirmation can be obtained.

Request a playback. You can ask the listener to repeat what you just said. It is reasonable to expect that the listener will paraphrase what you said. You can then verify if your message came across clearly.

Ask for a show of hands. This is an effective method to obtain confirmation for something that can be answered by Yes or No. You may follow through by asking one or two people in the audience why they raised their hand to indicate a No.

Open the floor to questions. Another way to verify that your audience understood what you just said is by pausing to take questions. You can judge the level of understanding of your audience by the quality of questions that they raise. Someone may even ask you to repeat or elaborate on an important point that you made.

Another method is a simple, straightforward question, "Do you have any concern on what I just said?" Of course, you have to use that technique discriminately as it may be offensive to your audience.

Finally, let me leave you with this thought: Communication is an integral part of life. We teach children to write and speak, but are we teaching them how to properly communicate?

Problem-solving

My bachelor's degree is in Industrial Engineering. We have a saying in Industrial Engineering that if you cannot define the problem, then there is no problem.

Thus, the first step in solving a problem is defining it. I have seen a few instances in my thirty plus years of work experience where technical specialists immediately performed troubleshooting without first talking to the person who reported the incident. The result was wasted time and effort. Therefore, the first step in solving a problem is asking this simple question, "What exactly is the problem?"

It is not enough to know that a customer is displeased. You have to find out what is making him or her frustrated or disappointed. It is not sufficient to say that "something is not working." You have to define what is not working.

Once you have defined the problem, the next step is to determine the root cause of the problem.

Determining the root cause of a problem involves listing several hypotheses and performing a process of elimination by testing each hypothesis one at a time.

Upon determining the root cause, you then need to develop alternative options to solve the problem. Are there approaches that were used to solve similar problems in the past? Do you need to consult an expert? Can the problem be solved by adding more equipment, people, or funding? Or is this a known problem that currently has no known solution?

The next step after you have listed alternative options to solve the problem is to decide on a solution. What helped me on many occasions was to draw what is called a *decision matrix*. A decision matrix is a simple table with three columns: 1) options, 2) advantages, and 3) disadvantages. You may then use the following factors to arrive at a decision: risk, cost, time, and resources or people.

Decision-making

A lot of people find decision-making difficult or intimidating. Deciding involves accepting responsibility for the decision. Whatever the outcome may be, people hold the one who made the decision accountable for the choice that was made. Rightly so.

However, it is important to underscore that the foundation of every decision is the value system of the person who makes the choice. In other words, every decision reflects the values of the person who calls the shots.

The truth of the matter is that we all make decisions. Every single day. We decide what we want to do for the day. We decide what to eat for breakfast, lunch, and dinner. We decide on what items we want to work on in the morning. We decide what we want to do after dinner.

One can argue that those are inconsequential compared to a decision that involves a multimillion-dollar investment. What we do not realize is, however trivial the decision may be, every choice we make reflects what we value in life. Every choice to do one thing means a decision not to do something else at that time.

Ideally, when you get to a point where you have to make big decisions, your values are clear to you. If not, then you may have difficulty making major decisions.

As you move up the corporate ladder, one decision you make can have a significant impact on the financial position of the company, while one choice you opt for may have a profound effect on many people.

As I stated in the previous section, you may use several factors to arrive at a decision: risk, cost, time, and resources or people.

Risk. Risk is an important factor to consider in making a critical decision. Most executives will view business matters from a risk perspective. What is the reputational risk if we choose Option A? What is the financial risk if we take Option B? Will Option C cause regulatory issues for the company? Will the company be potentially sued if Option D were to be taken? Will the company face a labor strike if Option E were chosen? What is non-negotiable is that risk to people is to be treated with utmost priority. For example, your response time to customers will slow down, but requiring your employees to stay at home during a pandemic will minimize their risk of contracting the disease.

Cost. It is important to manage the expenses of a company or even a household. The resources of a corporation or a family are finite. Therefore, cost is most often an important factor to consider when deciding. Vendor A sells exactly the same product as Vendor B; however, the former is giving a 15% discount. Consultant A, who has relatively the same level of experience as Consultant B, is quoting 10% less than his competitor.

Time. Time is a very precious commodity in life. There are only 24 hours in a day. That is all that every person is given. It is not negotiable. You cannot buy additional hours to extend your day. The term *spending time* is a profound one. It is like saying you have $86,400 a day in your life account that you cannot save or add to your account tomorrow. Once one second passes, you cannot take it back. It is spent. In business, speed is very important. You will hear people say, "Time is money." Therefore, time is often a significant factor in making business decisions. Builder A is slightly more expensive than

Builder B, but Builder A can finish the office building 45 days earlier than its competitor. That also means that we should have all the office furniture and equipment delivered immediately after the construction is complete. However, Supplier A cannot commit to that delivery date, while Supplier B can.

Resources or People. Human resources, in most cases, play a critical role in decision-making. Health, safety, and security of people always come first. Family commitments. Competency of workforce. Training needs. Mentoring. Job satisfaction. Motivating people. Salaries and benefits. Promotions. Reporting structures. Vacation time. Retirement. Resignations. Need for backup personnel. Hiring and firing. Consultants. People get tired. People get sick. People need time off. People need to be motivated. People need to be properly equipped to do their jobs. All these need to be considered in making decisions.

In summary, the decisions we make reflect what we value in life. We all make "small" decisions every day, but there are moments when we have to make major decisions. It is important for us to train our young people on how to make big decisions. How? We teach them values.

Motivating

Knowing how to motivate people is a critical skill in management. People who are motivated are known to perform much better than those who are not.

There are many things that motivate people: fame, fortune, social status, among others.

Abraham Maslow is widely known to be the author of *Maslow's Hierarchy of Needs.* Maslow's theory is that there are tiers of human needs: physical, safety, social (love and belonging), esteem, and self-actualization.

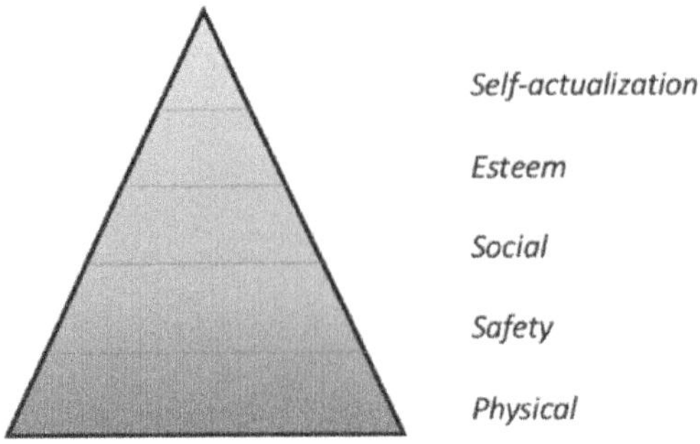

Maslow's Hierarchy of Needs

Physical needs cover water, food, shelter, and rest. Safety needs cover security and, of course, the feeling or assurance of being safe. Social needs cover friendships as well as intimate relationships. Esteem generally covers the feeling of accomplishment, prestige, and respect. Self-actualization is the highest level in the hierarchy of human needs. It involves achieving one's full potential, which may include the feeling or knowledge of being able to optimally use one's creative abilities.

Motivating a person requires a good understanding of that person. A person who is struggling with finances will respond well to a pay increase. However, someone who is already receiving a million-dollar paycheck may look for prestige and/or self-actualization. That said, you may still find an individual who is totally contented with where he is in life and does not strive for more.

Furthermore, the motivators that worked well for the generation of baby boomers may not necessarily be effective for today's millennials.

I had a conversation recently with a New York City high school teacher who told me that students nowadays are struggling to choose from a list of college courses. They are looking for courses that can give them a feeling of accomplishment or fulfillment. The mindset that one should take a course "that can pay the bills" does not appeal to a lot of millennials. It will be interesting to analyze many years from now how these young people navigated their careers.

In summary, you need to understand the psychology of an individual to be able to effectively motivate that person.

How are we teaching our young people to develop their motivational skills?

Delegation

Early in my career as a manager, I heard someone say, "If you are doing the work yourself, then you are not managing." In short, delegating tasks is an integral function of managers.

What is delegation?

Delegation is the assignment of any responsibility or authority to another person to carry out specific tasks or activities.

In my three decades as an IT leader, I noticed that people who are used to performing technical work tend to have difficulty transitioning to a supervisor or manager position. Why? Because they find it challenging to delegate tasks. Some say, "I was trained to do things myself." Others say, "I end up doing it myself because the outcome is not what I expected."

Clearly, a manager who heads a team of 10 people cannot and should not do the work of all of his people. If the manager tries to do that, then he becomes a bottleneck. A manager has to learn to delegate. Otherwise, he is not doing his work.

Similarly, a captain of a ship should not be spending days in the engine room doing repair and maintenance.

There are numerous reasons why a manager may hesitate to delegate. He may feel awkward delegating work to people who used to be his peers. Or he doesn't trust that the work will be done properly. Or perhaps he is used to doing the work himself.

What factors are involved in mastering the skill associated with delegation?

Delegation involves a healthy concept of one's self. Such an assessment of self allows you to appreciate the expertise of others and rely on their good judgment to successfully complete the work. No one is so great that he is the best in everything. No one. That is the truth.

Delegation involves trust. A manager who has difficulty trusting people will end up doing the work himself. On the other hand, delegation does not mean blind trust. A manager knows that things can go wrong and mitigates the risk. How?

Delegation involves giving complete and accurate instructions. Clearly, failure to provide proper directives will likely yield an undesired outcome. The person delegating the responsibility to another must ensure that the assignment is well understood and that the expected outcome is clear.

Delegation involves the ability to monitor progress. Once a task is assigned to someone, it is the responsibility of the manager to track the progress of that assigned person. While the expectation to deliver is on the assignee, the ultimate accountability rests on the manager who delegated the task. Monitoring, however, is not to be done in such a way that it impedes the progress of the work. When a manager assigns someone a task that needs to be completed in one hour, it is not reasonable to call the person every 10 minutes and spend three minutes each time to ask for an update.

Delegation involves mentoring. When a task is not performed as expected, the manager must sit down with the doer and analyze what happened. This is a critical step in order to draw lessons that can be used to help improve the performance of the assigned individual.

Delegation involves rewarding. When a task is done properly and the expected deliverable is met, it is important to recognize the doer. This reward may be in the form of cash, commendation, or a simple Thank You note. Such a gesture reinforces good behavior and motivates the assigned person.

Summary

The entrepreneurial skills and managerial capabilities mentioned in this chapter are by no means a complete list. Management books will be able to provide more details on skills an entrepreneur or a manager should possess.

What I would like to highlight in this chapter is that the implementation of automation in many industries will have a significant impact on the job security of laborers. The people who will likely survive this dramatic shift are the entrepreneurs and managers.

The key questions for this survival challenge, then, are as follows: How are we equipping our young people? Is the educational system sufficiently training our children on the necessary skills we have discussed? Are their subjects in school going to be relevant in the future? If not, then what should change? What areas of learning should be included in the core curriculum for elementary school, high school, and college?

Clearly, if the educational system is not addressing these questions, then it is imperative for parents to train their children on these core skills that will equip them to thrive in the future.

To my mind, parents are in the best position to teach their children how to develop entrepreneurial and managerial skills. Parents can share how they make their decisions, what factors they consider in decision-making, what mistakes they made in the past, and what lessons they learned.

Some of the most successful businesses that are still around today were passed on from one generation to another. That is only possible when one generation is committed to train the next and the latter is willing to learn.

The first concrete step for that to happen is for parents to reconnect with their children. The effort to reconnect involves both the parents and the children. Without that connection, it is unlikely that proper mentoring can take place.

Chapter 10
Analytical and Technical Skills

In the last few years, I have noticed that job ads for technical positions list "Must be creative" as one of the required skills. On the other hand, I have also noticed that job postings for artistic positions require the ability to analyze data.

People in the IT industry know that software development work requires creativity. User interface design, for example, demands a relatively good understanding of what is aesthetically pleasing to the eyes.

On the other hand, people in the advertising industry today know that data analysis plays a critical role in determining the success of a media campaign.

What this means is that more and more jobs in the future will require not only technical skills but also analytical skills. Clearly, AI and robotics will create new job opportunities involving both technical and analytical skills. However, these jobs will also require a certain degree of creative ability.

In a previous chapter, we discussed how to develop creative skills. In this chapter, let's talk about how to develop analytical skills.

What is analytical thinking?

Analytical thinking is the ability to break down a complex problem into small, manageable components.

In Chapter 9, we briefly discussed at a high level the problem-solving process relating to managerial skills.

However, analytical thinking is the underpinning skill that is foundational to problem-solving. There are many materials about analytical thinking and how to develop it. Allow me to bring it down to much simpler steps.

Assuming that you now have a good understanding of a problem confronting you, here are the basic steps in the analytical thinking process:

1. Data gathering
2. Organizing of data
3. Interpretation and analysis of data
4. Inference or conclusion

Data Gathering

Data gathering is a critical step in the analytical thinking process. Data can be obtained through various methods, including empirical observation, internet search, interviews, reading books or journals, sorting through video footages, system logs, among many others.

It is worth noting that not all data that you can gather may be relevant to the problem. Knowing what data is applicable to your problem is of primary importance. Why? Because you can waste a lot of time analyzing data that may be totally irrelevant to the issue you are trying to solve.

For example, when you search the internet for a specific topic, you may be presented with hundreds of links to a vast array of materials—news items, dictionaries, journals, blogs, videos, music, books, among many others. It is up to you to sift through the materials and determine what is pertinent to your research.

However, there may be instances when data that initially appears to be irrelevant to the problem may turn out to be material in helping you arrive at a solution.

More importantly, it is critical to determine if the source of the data is reliable. There are many materials on the internet that may appear to be from authoritative sources, but are actually based on hearsay or a hoax. Part of analytical thinking is being able to distinguish a trustworthy source of information from an unreliable one.

There is no cookie-cutter method in gathering data. Hands-on practice is the best teacher.

Organizing of Data

After gathering data, the next step in analytical thinking is organizing your data. There are various ways of doing this.

You can *group related data.* Logical grouping of data helps in analyzing facts that share the same context.

You can also *sort data.* Data can be sorted in sequential order, for example, or in order of occurrence. Event 1 happened on Monday, event 2 happened on Tuesday, event 3 happened on Thursday, and so forth and so on.

You can also organize data by *order of priority* or *order of importance.* If you are trying to find out why your sales for a specific product are dipping based on 550 reported complaints, it may be revealing to tally the complaints and list them in the order of priority. For example, the greatest number of complaints has to do with "defective product," followed by "no user manual."

You may *use visualization techniques* to help you organize your data. This could be a simple list, a pie chart, a bar graph, a heatmap, or a line graph.

I know the above examples are simplistic, but you get the picture.

Again, there are numerous books on analytical thinking, but there is no better way to learn it than to actually do it yourself.

Interpretation and Analysis of Data

Once you have organized your data, then you are ready for the next step, which is to interpret or analyze your data.

This is the step where you make sense out of the data that you have gathered and organized.

The key question to ask is this: What is the story that the data is telling?

Is the data showing a trend?

Is it an upward trend?

Is it a downward trend?

Is data showing a wave—meaning, up one year and then down the next?

Is the data telling a contradicting story? Why?

Is the data consistently supporting a single thesis? Or is it debunking a hypothesis?

Familiarity with the dataset is, of course, critical in the analysis process. Knowing why there are outliers or extremes is important. Understanding the basics of a few mathematical methods will also be helpful.

In case data is so disparate that what you deemed as "facts" do not appear to be telling any story, then it may be necessary to revalidate your source and method of gathering data, or reassess the validity of the data that you have been working with.

Inference or Conclusion

The final step—and perhaps the most important—in analytical thinking is drawing your conclusion.

In numerous cases, you will be able to draw an inference. In others, you may arrive at a conjecture that the data is not sufficient to draw a conclusion.

The conclusion may explain the root cause of a problem. In some instances, it can prompt a proposal to fund a project. In others, it may signal the reverse, meaning to stop or to defund a project.

In a hospital setting, a conclusion—most often a medical diagnosis—is very critical as it can result in the survival or death of a patient.

In managing a business, an inference may spell the success or failure of a product, a service, or even an entire company.

Conclusion

As discussed at the start of this chapter, more and more jobs in the future will require not only technical skills but analytical skills as well.

It is therefore imperative to train the young generation on how to develop their analytical skills.

The best way to teach them is to engage them in intelligent discussions that require analytical thinking.

Ask your children why they think the movie they have watched is "good." Analyze the behavior of the characters in the movie. Discuss what makes the plot or storyline interesting to them.

Ask them for suggestions for your next family vacation. Discuss the reasons for their recommendations.

Find out what they think of the government's decision on a specific matter. Ask them what they would do if they were the decision-makers.

Give concrete examples of how you dissected a problem, gathered data, organized data, analyzed data, and arrived at your conclusion. Tell them the good stories and the not-so-good stories—meaning, those that ended in failure because of a wrong conclusion.

They, too, can learn from your mistakes.

Chapter 11
Interpersonal Skills

We interact with people every day. The set of tools and techniques we use to communicate and interact with other people is called interpersonal skills. As such, it should not be a surprise that these skills of interaction are probably the most important of all human abilities. In the workplace, mastery of interpersonal skills is usually a key predictor of success.

While there will be social robots in the future, a gap will remain that only humans can fill. No company will send a robot to negotiate a multimillion-dollar project. Such deals are made with a firm handshake, eye contact, persuasion, and trust.

To be clear, interpersonal skills include both verbal and non-verbal communication skills. Understanding the nuances of a conversation, including body language, is all part of what we call interpersonal skills.

There are times when non-verbal signals play a critical role in sending a strong message to the other party. Whether subtle or obvious, such expressions, gestures, or body positions can show whether the person is angry, disappointed, excited, bored, listening, or not interested.

In the context of managing others, strong interpersonal skills are required in inspiring people, resolving conflict, building teamwork, promoting a positive work environment, mentoring and coaching, among many others.

As you can infer from the above, inspiring people cannot be effectively performed by a robot. The approach to motivating others will vary depending on the circumstances and the personalities involved. There is no AI algorithm that can surpass humans in this area.

A person is usually inspired by someone who personifies the goals that he or she is aiming for. An aspiring chef will be effectively motivated by a prominent restaurateur. Similarly, a young man who dreams of becoming a business owner will be inspired by a successful entrepreneur. Therefore, inspiring people is more than mere rhetoric; it is being a role model.

Conflict resolution is another area where robotic capabilities will not be able to surpass human skills. There are so many factors involved in resolving a conflict that there is no pre-defined set of questions that can be used to address the broad range of possible sticking points. Each conflict has to be treated in a way that is appropriate to the situation. Listening carefully to the nuances of what is being said is imperative. Body language is to be properly interpreted. Empathy is very important. Tone is calculated. Most importantly, trust has to be established.

Team building is another skill that is important when it comes to managing people. Teams are made up of individuals who come from different backgrounds. As such, rivalry and distrust can cause team members to lose focus on the goals of the group. Emotional baggage can slow down, if not destroy, the team. Therefore, team building requires understanding the strengths and weaknesses of every member, coaching and motivating each one, and rallying the whole team around a common goal. Again, there is no AI algorithm that can match human skills when it comes to team building.

Creating a positive work environment is an art. There may be a few guidelines that one can follow like expressing gratitude, respecting one another, and showing appreciation of co-workers. However, when and how these are done cannot be simply prescribed as a set formula. Collaboration and a positive attitude in the workplace do not magically happen. They are a product of determination and a laborious process that involves not just the leader but also the workers. As you can expect, there is no AI algorithm that can be superior to human ability when it comes to fostering a positive work environment.

Mentoring was briefly discussed in a prior chapter as a component of delegation. What is important to remember is that mentoring, as a form of human interaction, is founded on a trust relationship between a knowledgeable person and a less experienced one. Without trust, the mentoring process is bound to fail. That trust is based on the mentee's recognition of the mentor's knowledge and experience. Once again, this is an area where no AI algorithm can approximate that human-to-human exchange.

There are many aspects of interpersonal skills that were not discussed in this chapter. Nonetheless, a key takeaway is this: Interpersonal skills are not acquired by simply reading books or watching videos. They are best learned through hands-on experience.

This brings me to my last point.

I have been asked in some of my lectures if the traditional face-to-face education will survive the future. There may be many reasons why online training will be a norm in the coming years. However, one of the biggest benefits of in-person classes, regardless of age, is the development of interpersonal skills.

Technology can make remote communication more and more efficient. Be that as it may, there is no replacement for personal, face-to-face human interaction.

Chapter 12
Grit

It took them six years of research and repeated experimentation before the Wright brothers successfully created a flying machine.[109]

What is more fascinating, however, is that Wilbur Wright had been so affected by an injury in an ice hockey game that he went into depression, did not get his high school diploma, and cancelled his ambition to enter college.[110]

It is believed that Edison tried 10,000 times before he found the right material for his light bulb. When asked about his failed attempts, he said, "I have not failed. I have just found 9,999 ways that do not work."[111]

Grit.

Grit is defined as courage, perseverance, and strength of character. It means having a firm resolve to pursue something—stopping at nothing.

Grit is a disabled person vigorously training for months in preparation for a marathon. It is a kid born in a poor family and finishing top of the class in an Ivy League school. Grit is a father who relentlessly searches for a cure for his son's debilitating disease. It is a portrait of an aspiring concert pianist practicing eight hours every day since she was seven years old.

Grit starts with courage. Courage to be different. Courage to take on a challenge when others have given up. Courage to dream. And dream big.

Grit is being conscientious. It aims to "do it well," "do it thoroughly," and "do it right"—the first time (or many attempts later, if that's what it takes). Grit is laser-focused on achieving a goal.

Grit is endurance. It doesn't stop after one week of practice. Grit is willing to forego temporary gratification for a long-term goal. It doesn't look back at the six miles but focuses on the finish line that is 20 miles ahead. Grit is standing up immediately after a fall to finish the race.

Grit is stubborn resilience. It has an inner confidence drawn from optimism, perhaps even seen by others as careless optimism. It is not "I can do this" but "I *will* do this." It is not "I can finish" but "I *will* finish."

Finally, grit is the relentless pursuit of excellence. As they say in NASA, "Failure is not an option." Grit doesn't give up until it reaches a level of mastery. World class mastery. The best of the best.

Grit is what champions are made of. Grit distinguishes the outstanding from the mediocre.

In 1972, psychologist Professor Walter Mischel of Stanford University conducted what is now widely known as the Marshmallow Test.[112]

Professor Mischel's method was simple. A child was offered a choice of one marshmallow or pretzel to be enjoyed at that moment, or two marshmallows or pretzels if the child would be able to wait about 15 minutes for the treats. Years later, it was found that the children who waited longer in Mischel's experiment turned out to have better life outcomes.

While delayed gratification is one of the characteristics of grit, grit is much more than that. Ask people who you consider to be successful; most of them, if not all, will say that it took them many years of trying and failing before they attained success.

While most parents today want their children to succeed in life, the environment that the world has created for the young generation does not promote grit.

On-demand movies. On-demand music. Instant access to information. These all point to one thing: immediate gratification. You want it? You can have it right now. You don't have to wait.

It is not surprising, then, that young people today have less patience compared to teenagers in the 1960s. Teens today find waiting a pain. Perhaps not just the teens.

Brick and mortar department stores are now struggling to survive as more and more customers prefer online ordering. There is no queue to deal with. They can have items delivered right to their homes in a couple of days. Even groceries and restaurant food are now available online.

Ordering a cab is done with a mobile app. A few clicks and you will shortly have a cab waiting in front of your apartment.

While technology is clearly delivering efficiencies, it is also subliminally creating a mindset that waiting is a waste of time.

Here lies the challenge.

Waiting is an essential part of grit. Not the idle kind of waiting; but the willingness to go through a process, however arduous it may be, to achieve something great. It is understanding and accepting that mastering a skill takes time. It is knowing the value of patience, perseverance, resiliency, and endurance in order to pursue excellence.

If the world is not teaching grit to our young people, then it is up to parents and teachers to recalibrate their mindset.

It is OK for children to wait. It is OK for them to learn that they cannot have candy in between meals. It is OK for them to know that they have to work for something. It is OK for them to dream big for as long as they are willing to develop the skills needed to achieve that dream. It is OK for them to be bored and use their time for creative endeavors. It is OK for them to work on their assignments first before they are allowed to play their video games. It is OK for them to fail and learn from their mistakes.

The paradigm that each generation should be better than their parents is often misconstrued as more luxury, less work, and less failure for the young.

However, a generation that has less resilience, less patience, and less perseverance is a generation with a weak foundation. A building without a solid foundation will not withstand storms, earthquakes, or other types of stresses. Such a structure is bound to fail.

It is high time to rethink how we are preparing the next generation.

Let us not ignore the signs.

Chapter 13

The Value of Values

There have been countless discourses about how we have become more advanced in the last 30 years due to science and technology. However, there have been few discussions about values and virtues. It is not surprising that many people today consider values as old-fashioned and irrelevant. After all, what do robots and artificial intelligence have to do with values?

Machines do not have values. AI has no virtues. That said, values and virtues are among the key characteristics that distinguish us from machines. However, are values of any worth to mankind?

Humanity is entering what I call the *Age of Knowledge*. We are now exploring the possibility of a man-machine interface that will potentially increase our mental capacity and capabilities. We are figuring out our own intelligence in order to create machines that can think like us. We are making unprecedented advances in drug research, healthcare, science, and technology that will enable us to live longer.

As human knowledge increases, however, there will be a rise in new paradigms and an increase in conflicting perspectives between what is right and what is wrong. Norms, traditions, beliefs, and ethics will be challenged. What was once held as right will be argued as wrong. In the name of modern thinking, what was previously regarded as wrong will be debated as right. Time-tested values will be dismissed as old thinking.

Again, is there value in values?

In order to answer that question, allow me to give you an extreme example.

Imagine for a moment a world where everyone has no empathy for one another, no patience, no kindness, no ethics, no commitment to any ideals, no dependability, no regard for hard work, and no concern for life, liberty, truth, and justice. It is a portrait of a planet that is savage. It is a picture of a world where hatred, lies, egocentrism, thoughtlessness, and chaos are the norm. It depicts a barbaric community, not one you will call civilized.

That savage world will not emerge overnight.

Just as it takes decades to build up a generation that is better than the previous one, it will also take decades of desensitization, neglect, and tolerance to destroy the foundations of a generation.

As I said, I know the above is an extreme example. However, it drives home one important point: values are important as they set the boundaries for all of us to keep from driving humanity to extinction. Values regulate the behavior of individuals and communities. They are the guardrails that keep us from getting flung off-track and annihilating the human race from the face of the earth.

So, again, *is there value in having values?* Certainly!

However, I propose that there is an even more profound philosophical reason that underpins our values: *the purpose of our existence.*

It is innate for us humans to search for the essence of life, the meaning of our being here.

Machines will not inherently search for meaning or purpose. AI is simply a set of algorithms that run on computer hardware. However, we humans are different.

We have been trying to understand our origins. We have been attempting to explain the genesis of time, space, and the universe.

We try to make sense out of the pain and suffering that we experience. We examine what makes us happy. In search of meaning and fulfillment, some people pursue wealth or fame, others seek knowledge, nurture relationships, or practice the arts.

Even those who profess that life has no meaning have come to that conclusion because, to begin with, they themselves asked the question and searched for the answer. This underscores the point that we humans naturally look for the purpose of our existence.

"Why am I here?"

"Who am I?"

Our response to those questions will ultimately define our value system. Our value system will determine how we decide and how we behave. While it sounds simple and straightforward, we humans are a complex species. We are inherently a bunch of inconsistent beings, so that at times we act outside the bounds of our value system. Yet without a compass to guide us in our journey through life, we are more likely to get lost in the thick of the woods.

A person without a clear value system is like an individual who cannot distinguish north from south. While a value system is not a guarantee for perfection, a person who knows what he values will find his way back to the right track.

How does one's value system translate to the real world of society and mankind, in general?

The value system of business executives, for example, will translate into their company's mission, vision, organizational structure, management of people, hiring, budget, projects, operations, among many other things. On a much broader scale, the value system of the leaders of a nation will shape that country's laws and regulations; more importantly, it will have a significant impact on the lives of the entire population.

The critical question then becomes: What is the value system that is vital for us to teach the younger generation?

This is not a book on morals. However, let me share with you some of my personal insights.

To my mind, the value system that we need to teach the young generation is that which encourages empathy, care for self and others, kindness, respect, humility, patience, perseverance, industry, efficiency, dependability, reliability, commitment, excellence, consistency, honesty, learning from mistakes, forgiveness, generosity, hope, and love for God, country, and fellow humans.

While its history might have been marred by religious wars and human failings, the Christian faith teaches the value system I described above.

Again, this is not a book on morals nor is it a book on theology.

Let me leave you with this thought: If the predictions of prominent AI writers will prove to be true, then the future will shake the foundations of human character. The values that we teach the young generation today can spell their survival or downfall in the years to come.

Chapter 14

Reimagining the Future

As I write this final chapter, the world is on its sixth month of a devastating global pandemic brought about by a novel coronavirus. Just like most people in the US, my family and I have been in a mandatory lockdown inside our home for some months now, following state regulations.

The virus has now reached 211 countries and has infected close to ten (10) million people across the globe. This novel coronavirus has claimed over 500,000 lives worldwide—just under six months since it was first reported to the World Health Organization on December 31, 2019. In the US alone, this highly infectious microorganism has already caused the death of over 128,000 people.[113]

Companies around the world have moved to telework, while schools have turned to remote learning.

Due to repeated warnings from health officials regarding a possible second wave in Fall 2020, there have been talks of a "new normal" involving social distancing, wearing masks in public places, controlled face-to-face gatherings, continuation of telework, multi-modal teaching and learning (in-person mixed with online), among many others.

In a race to find a cure and a vaccine, technology and drug companies have turned to AI. IBM used its fastest supercomputer called Summit to run thousands of simulations in order to analyze which drug compounds might effectively stop the coronavirus from infecting host cells. The IBM high performance computer identified 77 drug compounds.[114]

Infervision's Coronavirus AI Solution was used during the epidemic outbreak at Tongji Hospital in Wuhan. Infervision AI expedited CT diagnosis by way of pneumonia marking, abnormal and severe case analysis, patient triage, medical resources coordination, prior case comparisons, and treatment assessments.[115]

On April 3, 2020, Bloomberg reported that Google was working with European researchers to trace the spread of the coronavirus. Here is an excerpt from that article:

> Google is working with researchers in Europe to track the spread of the coronavirus using troves of location data gathered from smartphones.
> The search engine giant is collaborating with academics from the University of Southampton in the U.K., who in turn are working with the European Centre for Disease Prevention and Control, according to several people involved in the project.
> The location data, which Google collects from location-enabled apps such as Google Maps, has been shared with the researchers in an aggregated and anonymized format. It can't be used to track an individual person; rather, it shows broad patterns of movement across entire countries over periods of time, according to the people.[116]

Robotics also played a key role in addressing the pandemic. Some hospitals in China were equipped with robots to clean and disinfect hospital rooms and working areas, deliver food to patients, and measure their temperature. CloudMinds, a Beijing-based robotics company, sent 14 robots to Wuhan, China to help with patient care during the coronavirus crisis.[117]

Furthermore, a police robot was deployed to patrol areas in the Tunisian capital of Tunis.[118]

One of the many stark and harsh realizations that will come out of this pandemic is that humans are vulnerable to sickness.

The undesired result is that companies will likely turn to disease-immune robots and AI to perform work that used to be dominated by humans.

Warehouse automation, self-driving trucks, and delivery robots will play a critical role in ensuring continuity of service in the logistics industry. Why is this important? People will turn to online ordering of groceries and other essentials. So, large grocery chains will deploy robots to run their online order fulfilment service.

More and more hospitals will look into robots that can perform critical functions such as disinfecting hospital rooms, delivering food and medicines to patients, and even tasks usually assigned to nurses.

Hospitals will implement AI solutions that can speed up diagnosis. This can range from AI-enabled radiologic image analysis all the way to patient flow in the hospital.

Drug companies will incorporate AI in their research projects to expedite drug discovery.

Countries will make use of contact tracing technologies using mobile phone location information to help control the spread of infectious diseases in the future.

Law enforcement agencies will employ robots and drones to enforce social distancing regulations.

In other words, the world will be forced to reimagine the future. The new normal will continue to evolve and will touch the life of every individual on this planet. That new normal will involve robots, AI, and other forms of automation.

The downside of this path is that it can cause the loss of millions of jobs. Governments will need to step up in order to keep the economy going. Universal Base Income for all citizens and residents may be one of the options that will be placed on the table.

People living in crowded cities will naturally rethink their priorities. Mass migration to less populated states is more probable than possible. Companies will need to be more open to telework in order to keep talent.

Mental health and well-being of millions of people will be affected one way or another. Clearly, the social life of many will continue to be disrupted.

Mass transit, cruises, air travel, vacation getaways, schooling options, concert going, sports events, movie viewing, restaurant dining, church services, among many others will have to be reimagined.

This also means that investments and business ideas will evolve. Outsourcing to other countries will need serious rethinking. Trade agreements, production, shipping, freight train operations, port operations, and many others will have to be revisited.

To my mind, the coronavirus pandemic is a catalyst that is accelerating the transformation that a number of AI writers have been predicting for a long while now. AI and robotics will be an integral part of life in the years to come. They will help solve some of the most difficult challenges that humanity is facing and will yet encounter. At the same time, these very technologies can and will take over the jobs of many.

In the final analysis, whatever turn the future may take, the responsibility rests on each one of us to prepare ourselves—and the next generation as well.

The future will not be status quo. It will not be business as usual. Whatever formula of success worked in the last three decades will not be fully applicable in the next 30 years.

One thing is for sure.

Throughout the history of mankind, success favored the ones who were ahead of their time. The thought leaders. The visionaries. The pioneers. The explorers. The founders. The inventors. The originals. The prime movers. The change agents. The early adopters. They who had the unique ability…to reimagine the future!

INDEX

NOTES

Introduction

[1] Link to a DeepMind website: https://deepmind.com/; link to a DeepMind blog on AlphaGo: https://deepmind.com/blog/article/alphago-zero-starting-scratch

[2] Source of information: https://senseis.xmp.net/?NumberOfPossibleGoGames

[3] Source of the information on Lee Se-dol's retirement: https://www.bbc.com/news/technology-50573071

[4] Link to IBM Watson's website: https://www.ibm.com/watson?p1=Search&p4=p50370936620&p5=e&cm_mmc=Search_Google-_-1S_1S-_-WW_NA-_-ibm%20watson_e&cm_mmca7=71700000060771776&cm_mmca8=aud-384354108630:kwd-2846208261&cm_mmca9=EAIaIQobChMIjcqO8NWS5wIVApyzCh3ZcgpKEAAYASAAEgLkN_D_BwE&cm_mmca10=406603549803&cm_mmca11=e&gclid=EAIaIQobChMIjcqO8NWS5wIVApyzCh3ZcgpKEAAYASAAEgLkN_D_BwE&gclsrc=aw.ds

[5] Link to the website of *Jeopardy!* https://www.jeopardy.com/

[6] Link to an Oxford Economics article entitled "How Robots Change the World": https://www.oxfordeconomics.com/recent-releases/how-robots-change-the-world

[7] Link to the November 19, 2019 article entitled "Artificial Intelligence Will Obliterate These Jobs by 2030" released by Fortune.com that cited the Forrester predictions: https://fortune.com/2019/11/19/artificial-intelligence-will-obliterate-these-jobs-by-2030/

[8] Link to the Babylon Health website: https://www.babylonhealth.com/

[9] Opencog is a project that aims to build an open source framework for Artificial General Intelligence. Here is a link to a webpage containing information on Opencog: https://wiki.opencog.org/w/The_Open_Cognition_Project

Chapter 1: The Future of Transportation

[10] Link to AutoX's website: https://www.autox.ai/

[11] Link to The Sunday Times article entitled "AUTONOMOUS TAXIS HAVE MADE THEIR DRIVERLESS DEBUT IN LONDON": https://www.driving.co.uk/news/autonomous-taxis-made-driverless-debut-london/

[12] Link to The Sunday Times article: https://www.driving.co.uk/news/autonomous-taxis-made-driverless-debut-london/

[13] Link to the Digital Trends article: https://www.digitaltrends.com/cars/gm-requests-green-light-to-ditch-steering-wheel-in-its-self-driving-cars/

[14] Link to the Venture Beat article: https://venturebeat.com/2019/12/31/baidu-secures-licenses-to-test-self-driving-cars-in-beijing/

[15] Link to the Goldman Sachs article: https://www.goldmansachs.com/insights/pages/the-car-industry-shifts-gears.html

[16] Source of information: https://www.britannica.com/technology/ship/17th-century-developments

[17] Source of information: https://www.vice.com/en_us/article/z43xg3/automated-ports-have-dockworkers-in-the-netherlands-threatening-strikes

[18] Link to The Guardian article published on April 7, 2016: https://www.theguardian.com/technology/2016/apr/07/convoy-self-driving-trucks-completes-first-european-cross-border-trip

[19] Link to Traton's website: https://traton.com/en/company.html

[20] Link to Traton's news article: https://traton.com/en/newsroom/stories/Scania-and-Rio-Tinto-trialling-autonomous-truck-.html

[21] Link to the NBC News article: https://www.nbcnews.com/mach/science/uber-just-unveiled-prototype-its-futuristic-air-taxi-ncna872771

[22] Link to the USA Today article: https://www.usatoday.com/story/money/cars/2018/07/17/aston-martins-flying-car-concept-fit-james-bond/792779002/

[23] Link to the Strait Times article: https://www.straitstimes.com/multimedia/flying-taxis-could-soon-take-off-in-singapore

[24] Link to The Verge article: https://www.theverge.com/2019/11/1/20943407/ehang-flying-taxi-drone-ipo-initial-offering-100-million

[25] Link to a February 8, 2020 Washington Post news article regarding Kobe Bryant's helicopter crash: https://www.washingtonpost.com/local/trafficandcommuting/kobe-bryant-helicopter-crash-underscores-industrys-long-running-safety-struggles/2020/02/08/42a1aba6-4444-11ea-b503-2b077c436617_story.html

[26] Link to Smithsonian Magazine article entitled "WORLD WAR I: 100 YEARS LATER, Unmanned Drones Have Been Around Since World War I": https://www.smithsonianmag.com/arts-culture/unmanned-drones-have-been-around-since-world-war-i-16055939/

[27] Link to Amazon Prime Air website: https://www.amazon.com/Amazon-Prime-Air/b?ie=UTF8&node=8037720011

[28] Link to the Futurism webpage: https://futurism.com/the-byte/ben-jerrys-ice-cream-drone-delivery-system

Chapter 2: The Future of Healthcare

[29] Link to the CB Insights article: https://www.cbinsights.com/research/artificial-intelligence-deals-tracker-heatmap/

[30] Link to Pixel Scientia Labs website: http://www.pixelscientia.com/pathology.html

[31] Link to Proscia's website: https://proscia.com/

[32] Link to Paige.AI's website: https://paige.ai/

[33] Link to the AACR article entitled "Startups Bring AI to Pathology": https://cancerdiscovery.aacrjournals.org/content/early/2018/02/23/2159-8290.CD-NB2018-019

[34] Link to the Health Imaging article: https://www.healthimaging.com/topics/artificial-intelligence/2018-investments-ai-medical-imaging-startups

[35] Link to Intuitive's website: https://www.intuitive.com/en-us

[36] The December 31, 2018 article that appeared on RBR's website: https://www.roboticsbusinessreview.com/health-medical/5-surgical-robots-2019/

[37] Source of information: Scientific American website. https://www.scientificamerican.com/article/cost-to-develop-new-pharmaceutical-drug-now-exceeds-2-5b/

[38] Source of information: Cancer Research UK website. https://www.cancerresearchuk.org/find-a-clinical-trial/how-clinical-trials-are-planned-and-organised/how-long-it-takes-for-a-new-drug-to-go-through-clinical-trials

---

[39] Link to the Pharmaceutical Technology's May 25, 2018 article: https://www.pharmaceutical-technology.com/comment/artificial-intelligence-future-drug-discovery/

[40] Link to MIT News article on a new antibiotic compound: http://news.mit.edu/2020/artificial-intelligence-identifies-new-antibiotic-0220

[41] Link to Babylon Health's website: https://www.babylonhealth.com/

[42] Link to the TechCrunch article on Babylon Health: https://techcrunch.com/2019/08/02/babylon-health-confirms-550m-raise-to-expand-its-ai-based-health-services-to-the-us-and-asia/

[43] HealthTap's website: https://www.healthtap.com/

[44] Sense.ly's website: www.sense.ly

[45] Push Doctor's website: https://www.pushdoctor.co.uk/

[46] Your.MD's website: https://www.your.md/

[47] Doctor Care Anywhere's website: https://doctorcareanywhere.com/

[48] Google is a multibillion-dollar global technology company that specializes in internet search. Here is Google's website: https://www.google.com/

[49] Link to the September 18, 2019 BBC News article entitled "Google swallows DeepMind Health": https://www.bbc.com/news/technology-49740095

[50] Bloomberg is a private financial, data, software, and media company based in Manhattan. Here is a link to the Bloomberg website: https://www.bloomberg.com/

[51] Link to the June 19, 2019 Bloomberg article: https://www.bloomberg.com/press-releases/2019-06-19/first-ever-successful-mind-controlled-robotic-arm-without-brain-implants-jx3jrpv3

[52] Link to the Open Bionics website: https://openbionics.com/

[53] Here is the Access and Mobility Professional website: https://www.accessandmobilityprofessional.com/

[54] Link to the Access and Mobility Professional article: https://www.accessandmobilityprofessional.com/prosthetics-firm-targets-nhs-affordable-3d-printed-super-hero-arms/

[55] Link to the Johns Hopkins Applied Physics Laboratory website: https://www.jhuapl.edu/

[56] Link to the Futurism article on smart stethoscope: https://futurism.com/the-byte/smart-stethoscope-diagnose-pneumonia

[57] Link to the Health Care Originals website: http://healthcareoriginals.com/

[58] Link to Bigfoot Biomedical's website: https://www.bigfootbiomedical.com/

Chapter 3: Robotics

[59] Link to the Bigfoot Biomedical article: https://www.bigfootbiomedical.com/series-c-financing-led-by-abbott

[60] Source of information: https://interestingengineering.com/15-engineers-and-their-inventions-that-defined-robotics

[61] Link to the website of Kawasaki Robotics: https://robotics.kawasaki.com/en1/?utm_source=google&utm_medium=cpc&utm_campaign=Robotics&utm_term=Homepage&gclid=EAIaIQobChMIqbyBsI-R5wIVmZOzCh0lpwU3EAAYASAAEgI6yfD_BwE

[62] Source of information on Kawasaki Robotics: https://robotics.kawasaki.com/en1/

[63] Link to Mitsubishi Electric's website: https://www.mitsubishielectric.com/fa/products/rbt/robot/

[64] Link to the website of Fanuc Robotics: https://www.fanucamerica.com/products/robots

[65] Link to the website of Epson Robots: https://epson.com/industrial-robots-factory-automation?utm_medium=cpc&utm_source=google&utm_campaign=Brand%2BProduct+Category%2BRobots%2BUS%2BENG%2BSPART&utm_adgroup=Robot+-+General%2BEXACT&utm_term=Epson+robots&gclid=EAIaIQobChMImen7mZaR5wIVjJOzCh3tzQFaEAAYASAAEgLR4fD_BwE&gclsrc=aw.ds

[66] Link to the website of Comau Robotics: https://www.comau.com/en

[67] Link to Durr's website: https://www.durr.com/en/products/paint-shop-application-technology/paint-robots-paint-machines/paint-robots-automotive/

[68] Link to the website of Universal Robots: https://www.universal-robots.com/q1us/find-your-way-through-automation/?utm_source=Kragen&utm_medium=ppc&utm_campaign=HQ5things&utm_content=whitepaper&gclid=EAIaIQobChMIlYvU_5aR5wIVCY6zCh14zwepEAAYASAAEgKaD_D_BwE

[69] Source of list of top robotics companies: https://blog.technavio.com/blog/top-21-companies-in-the-industrial-robotics-market

[70] Link to the November 10, 2017 Alltech article: https://www.alltech.com/blog/flocking-digital-future-poultry-technology

[71] Link to Sintef's website: https://www.sintef.no/en/latest-news/using-robots-to-get-more-food-from-raw-materials/

[72] The Los Angeles Times also known as the LA Times is a daily newspaper published in Los Angeles, California. Here is a link to their website: https://www.latimes.com/

[73] Fox News is an American news network. Here is a link to their website: https://www.foxnews.com/

[74] Link to the Fox News article: https://www.foxnews.com/tech/a-japanese-hotel-is-going-to-swap-robots-for-humans

[75] Link to the Fox News article: https://www.foxnews.com/tech/a-japanese-hotel-is-going-to-swap-robots-for-humans

[76] Business Insider is a financial business news website. Here is a link to their website: https://www.businessinsider.com/

[77] Link to the Business Insider article: https://www.businessinsider.com/alibaba-hotel-of-the-future-robots-ai-2019-10

[78] Source of information: https://www.socialtables.com/blog/hospitality-technology/hotel-brands-robot/

[79] Source of information: https://online-engineering.case.edu/blog/medical-robots-making-a-difference

[80] Link to The Medical Futurist website: https://medicalfuturist.com/9-exciting-medical-robot-facts/

[81] Aethon is a medical robots manufacturing company based in Pittsburgh, Pennsylvania.

[82] Source of information: https://www.roboticsbusinessreview.com/health-medical/blue_collar_delivery_robots_making_gains_in_u-s-_hospitals/

[83] Source of information on Akara Robotics: https://time.com/5825860/coronavirus-robot/?utm_source=feedburner&utm_medium=feed&utm_campaign=Feed%3A+time%2Ftopstories+%28TIME%3A+Top+Stories%29

[84] ZDNet is a website that provides business technology news. Here is a link to their website: https://www.zdnet.com/

[85] Link to the ZDNet article dated October 11, 2019: https://www.zdnet.com/article/nurse-robot-is-now-making-the-rounds-at-major-hospitals/

[86] Link to the Statista article: https://www.statista.com/statistics/241494/median-age-of-the-us-population/

[87] Link to the Bloomberg article: https://www.bloomberg.com/news/articles/2019-06-20/half-of-americans-are-now-over-the-age-of-38-census-data-show

[88] Link to the Reuters article on eldercare robots in Japan: https://www.reuters.com/article/us-japan-ageing-robots-widerimage/aging-japan-robots-may-have-role-in-future-of-elder-care-idUSKBN1H33AB

[89] Link to the Synced article: https://syncedreview.com/2019/06/01/grandmas-robot-how-ai-is-revolutionizing-elder-care/

[90] Source of information: https://www.hansonrobotics.com/david-hanson/

[91] Link to WBUR's article on Spot: https://www.wbur.org/news/2019/11/25/boston-dynamics-robot-dog-massachusetts-state-police

[92] Link to the Bloomberg article on Spot: https://www.bloomberg.com/news/articles/2020-02-11/spot-the-dog-is-oil-producer-aker-bp-s-first-robotic-employee

[93] Link to Boston Dynamics's website on Atlas: https://www.bostondynamics.com/atlas

[94] Link to the *Forbes* article: https://www.forbes.com/sites/simonchandler/2020/01/20/how-artificial-intelligence-will-make-decisions-in-tomorrows-wars/#5c05e6ee4e01

Chapter 4: AI in Finance

[95] Source of information on Epoque: https://builtin.com/artificial-intelligence/ai-trading-stock-market-tech

[96] Source of information: https://builtin.com/artificial-intelligence/ai-trading-stock-market-tech

[97] Source of information: https://builtin.com/artificial-intelligence/ai-trading-stock-market-tech

[98] Link to Welcome.AI website: https://www.welcome.ai/arbitrage-technologies

[99] Link to VantagePoint Software's website: https://www.vantagepointsoftware.com/software/

[100] Allen Zhang is the founder of WeChat. Source: https://en.wikipedia.org/wiki/Allen_Zhang

[101] Source of information: https://www.forbes.com/sites/jackkelly/2019/10/08/wells-fargo-predicts-that-robots-will-steal-200000-banking-jobs-within-the-next-10-years/#20b85b7568d7

[102] Source of information: https://www.forbes.com/sites/jackkelly/2019/12/10/artificial-intelligence-is-superseding-well-paying-wall-street-jobs/#240986f0524d

[103] Source of information: https://www.technologyreview.com/f/614469/france-plans-to-use-facial-recognition-to-let-citizens-access-government-services/

[104] Source of information: https://www.bbc.com/news/uk-politics-51550421

[105] Source of information: https://www.cnbc.com/2018/07/17/artificial-intelligence-to-create-more-jobs-than-it-destroys-pwc-says.html

[106] Link to the *Forbes* article entitled "Artificial Intelligence To Create 58 Million New Jobs By 2022, Says Report" dated September 19, 2018: https://www.forbes.com/sites/amitchowdhry/2018/09/18/artificial-intelligence-to-create-58-million-new-jobs-by-2022-says-report/#7b5879914d4b

Chapter 7: Creativity and Connecting the Dots

[107] Source of definition:
https://www.coursehero.com/file/p4q2rdd/Creativity-by-definition-is-the-use-of-imagination-or-original-ideas-especially/

[108] Link to the Merriam Webster dictionary definition of empathy:
https://www.merriam-webster.com/dictionary/empathy

Chapter 12: Grit

[109] Source of information on the Wright brothers:
https://airandspace.si.edu/exhibitions/wright-brothers/online/fly/

[110] Source of information on Wilbur Wright:
https://www.history.com/topics/inventions/wright-brothers

[111] Source of information on Thomas Edison:
https://www.businessinsider.com/thomas-edison-and-michael-jordan-were-failures-2010-9

[112] Source of information:
https://www.apa.org/helpcenter/willpower-gratification.pdf

Chapter 14: Reimagining the Future

[113] Source of data (as of May 19, 2020):
https://www.cnn.com/interactive/2020/health/coronavirus-maps-and-cases/

[114] Source of information:
https://www.cnn.com/2020/03/19/us/fastest-supercomputer-coronavirus-scn-trnd/index.html

[115] Source of information:
https://www.itnonline.com/content/infervision-frontlines-against-coronavirus

[116] The April 3, 2020 article from Bloomberg: https://www.bloomberg.com/news/articles/2020-04-03/google-joins-with-u-k-researchers-to-track-coronavirus-cases

[117] CNBC article on the use of robots in hospitals in China: https://www.cnbc.com/2020/03/23/video-hospital-in-china-where-covid-19-patients-treated-by-robots.html

[118] BBC news on a police robot deployed in Tunis: https://www.bbc.com/news/world-africa-52148639

About the Author

Elzar Simon has more than thirty years of experience as an information technology global thought leader. He has lived in Australia, North America, and Asia and has worked as an IT strategist in a broad range of industries, including finance, logistics, health care, government, IT service, and higher education. He currently works as senior IT director for global infrastructure in one of the top universities in the Northeast United States.

Made in the USA
Monee, IL
12 July 2020

35577838R00100